ISBN 978-1-332-89621-9
PIBN 10434309

1 MONTH OF FREE READING

at

www.ForgottenBooks.com

By purchasing this book you are eligible for one month membership to ForgottenBooks.com, giving you unlimited access to our entire collection of over 1,000,000 titles via our web site and mobile apps.

To claim your free month visit:

www.forgottenbooks.com/free434309

English
Français
Deutsche
Italiano
Español
Português

www.forgottenbooks.com

Mythology Photography **Fiction**
Fishing Christianity **Art** Cooking
Essays Buddhism Freemasonry
Medicine **Biology** Music **Ancient**
Egypt Evolution Carpentry Physics
Dance Geology **Mathematics** Fitness
Shakespeare **Folklore** Yoga Marketing
Confidence Immortality Biographies
Poetry **Psychology** Witchcraft
Electronics Chemistry History **Law**
Accounting **Philosophy** Anthropology
Alchemy Drama Quantum Mechanics
Atheism Sexual Health **Ancient History**
Entrepreneurship Languages Sport
Paleontology Needlework Islam
Metaphysics Investment Archaeology
Parenting Statistics Criminology
Motivational

SMITH AND DUKE'S

AMERICAN

STATISTICAL ARITHMETIC;

DESIGNED FOR

ACADEMIES AND SCHOOLS.

BY

FRANCIS H. SMITH, A.M.

SUPERINTENDENT AND PROFESSOR OF MATHEMATICS IN THE VIRGINIA
MILITARY INSTITUTE; LATE PROF. OF MATHEMATICS IN HAMPDEN
SIDNEY COLLEGE, AND FORMERLY ASSISTANT PROF. IN
THE U. S. MILITARY ACADEMY, WEST POINT.

AND

R. T. W. DUKE,

ASSISTANT PROFESSOR OF MATHEMATICS IN VIRGINIA
MILITARY INSTITUTE.

THIRD EDITION.

PHILADELPHIA:
THOMAS, COWPERTHWAIT & CO.
1845.

STEREOTYPED BY J. FAGAN.

(2)

PREFACE.

THE characteristic feature of the following Treatise is simple but peculiar. It has been the design of the author to illustrate the various rules of Arithmetic by examples selected, as far as practicable, from the most prominent facts connected with the History, Geography, and Statistics of our country. Multiplication of simple numbers is illustrated by such examples as the following :—

"The Congress of the United States is composed of a Senate, consisting of 2 members from each of the 26 States, and a House of Representatives of 242 members; each member receives $8 per day during the session of Congress: what is the daily expense to the Government for the pay of the members?"

Here is a simple example in multiplication, and yet what valuable information does it convey! The constitution of Congress — the number of States — the equality of their representation in the Senate — the number of members in the House of Representatives — their pay, &c. How many citizens are there who exercise the elective franchise who are ignorant of these facts!

It will readily appear, how, by extending this principle so as to embrace questions relating to the Commerce, Manufactures, Agriculture, &c. of the several States, the interest in the study of this essential branch of human knowledge may be increased. Arithmetic thus becomes a medium for communicating much important information, which will be readily apprehended by the youthful mind, and impressed upon it through life. If the study of the *American Statistical Arithmetic* shall be an instrument in imparting to the American youth, in any degree, an appreciation of the greatness and resources of their country, and of the bounty of Providence towards it, the labour of its author will not have been in vain.

Great care has been bestowed upon the scientific portion of the work, so as to make the rules simple and intelligible to the least advanced scholar; those explanations being given which an experience of many years in teaching has shown to be the best.

A correct idea of a *unit* being indispensable to a thorough knowledge of arithmetic, the attention of the pupil is prominently directed to the necessity of some standard in comparing quantities, and to the various kinds which must be used to correspond with the different kinds of quantity which are compared. The definition and uses of the *unit* being the same in all the branches of mathematics, an accurate perception of the subject in arithmetic will greatly facilitate the acquisition of the higher branches.

The division of numbers into *Simple* and *Compound* has been adhered to. The term *Simple* is applied to those numbers which are expressed in terms of the *same unit*. Thus, £3, £5, and £10 are *simple* numbers, the unit in each being the £; and the operations of arithmetic may be performed upon them in the same manner as if the kind of unit were not expressed. The number £2 3s. 6d. is a *compound* number, being compounded of units of different kinds — viz. pounds, shillings, and pence.

An Appendix, embracing *Mensuration*, with its applications to *Carpentry*, *Masonry*, *Land Surveying*, &c., and a brief exhibition of the two methods of *Book-Keeping* by *single* and *double entry*, will add much to the value of the work.

A Key for the use of teachers will accompany the arithmetic.

To prevent any mistake by confounding this work with the arithmetic published by *Mr. Roswell C. Smith*, the author has associated with his own, the name of his assistant, Mr. R. T. W. Duke, for whose kindness in- aiding him in the revision of the proofs, he would take this occasion to express his acknowledgments.

FRANCIS H. SMITH.

Virginia Military Institute,
  January 1, 1845.

TABLE OF CONTENTS.

PROPORTION.

REDUCTION OF CURRENCIES.

INTEREST.

ANNUITIES.

COMMISSION AND INSURANCE.

DISCOUNT.

PROFIT AND LOSS.

EQUATION OF PAYMENTS.

DUODECIMALS.

FORMATION OF POWERS.

EXTRACTION OF THE SQUARE ROOT.

EXTRACTION OF THE CUBE ROOT.

ARITHMETICAL PROGRESSION.

GEOMETRICAL PROGRESSION.

APPLICATION OF ARITHMETIC TO MENSURATION.

APPLICATIONS OF MENSURATION.

ARITHMETIC.

DEFINITIONS — NUMERATION.

1. QUANTITY is anything that will admit of increase or diminution. *Time, space, weight,* and *motion,* are quantities.

MATHEMATICS is the science of quantity.

Questions. What is quantity? Give some examples of quantity. What is mathematics?

2. An UNIT *is a single quantity which is used to compare quantities of the same kind with each other.* NUMBERS *express how many of these units are considered.*

Thus—if we say a box contains ten pounds of tea, *one pound of tea* is the *unit,* and *ten* the *number.* In six yards of cloth, *one yard* of cloth is the unit, and six the number; in twenty dollars, *one dollar* is the unit, and twenty the number. The amount of tea, cloth, and money, is in each case compared with its own unit, which is one pound of tea for the first, one yard of cloth for the second, and one dollar for the third.

Q. What is a unit? What do numbers express? In ten pounds of tea, what is the unit? What is the number? In six yards of cloth? In twenty dollars? In fifty miles? In each case what is the amount of tea, cloth, &c., compared with?

3. ARITHMETIC, which is a branch of Mathematics, treats of quantity when it is expressed by numbers. *Arithmetic then is the science of numbers.*

Numbers are expressed by certain characters which are called FIGURES. Only *ten* of these characters are used. They are,

0 which is called zero, a cipher, or Naught,
1 - - - - - - - - - - - One,
2 - - - - - - - - - - - Two,
3 - - - - - - - - - - - Three,
4 - - - - - - - - - - - Four,
5 - - - - - - - - - - - Five,
6 - - - - - - - - - - - Six,
7 - - - - - - - - - - - Seven,
8 - - - - - - - - - - - Eight,
9 - - - - - - - - - - - Nine.

The nine figures 1, 2, 3, 4, 5, 6, 7, 8, 9, are called *digits* or *significant figures*.

Q. What is arithmetic ? Of what does it treat ? Of what science is it a branch ? How are numbers expressed ? How many figures are there ? Write them upon your slate. Which of these figures are called digits ? What else are they called ?

4. Although every number might be represented by a distinct figure, we have no separate character to express a number greater than *nine*. We express numbers greater than nine, by combining the characters already known.

Q. Have we a separate character for numbers greater than nine ? Why not ? How do we express such numbers ?

5. To express *ten*, we use the two characters 1 and 0, placing the 0 on the right of the 1. Thus 10, which is ten. The position occupied by the figure on the right is called the *units' place*, that by the figure on the left the *tens' place*. The 0 is in the units' place, the 1 in the tens' place. The same figure is therefore ten times greater in the tens' place than in the units' place.

In like manner, we can express two tens, or twenty, three tens, or thirty, &c., as far as nine tens, or ninety, by placing 0 on the right of 2, 3, 4, 5, &c.

Thus,   20 which is Twenty,
            30 - - - Thirty,
            40 - - - Forty,
            50 - - - Fifty,
            60 - - - Sixty,
            70 - - - Seventy,
            80 - - - Eighty,
            90 - - - Ninety.

Q. How is ten expressed? Which figure occupies the units' place in 10? Which the tens' place? How much greater is the same figure in the tens' place than in the units' place? How is twenty expressed? Thirty? Forty? Fifty? &c. Which figure occupies the tens' place in 20? Which the units' place? In 30? In 60? In 90? Write these numbers upon your slate.

6. To express the intermediate numbers between 10 and 20, 20 and 30, &c., we place in the tens' and units' places, the number of tens and units of which the number is composed. Thus,

Eighteen contains 1 ten and 8 units, and is written   18
Twenty-five contains 2 tens and 5 units - - - - 25
Thirty-seven contains 3 tens and 7 units   - - - 37
Forty-four contains 4 tens and 4 units   - - - - 44
Ninety-nine contains 9 tens and 9 units - - - - 99

Ninety-nine is the greatest number which is expressed by two figures.

Q. How are the intermediate numbers between 10 and 20 expressed? How many tens and units in eighteen? How is it written? How many tens and units in forty-four? Write this number? How many tens and units in fifty-six? In seventy-two? In eighty-eight? In ninety-nine? What is the greatest number expressed by two figures?

7. To express *one hundred*, we place two *ciphers* on the right of 1. Thus 100, which is *one hundred*. The place occupied by the 1 in this case is called the *hundreds' place*. *It is ten times greater in the hundreds' place than in the tens' place, and one hundred times greater than in the units' place.*

In like manner, 200, 300, 400, 900, represent *two hundred, three hundred, four hundred,* &c.

We may express any intermediate number by writing down the number of *hundreds, tens,* and *units,* of which it is composed, placing the hundreds in the hundreds' place, the tens in the tens' place, and the units in the units' place. Thus,

One hundred and fifty-seven, contains 1 hundred,
  5 tens, and 7 units, and is written - - - - - 157
Three hundred and sixty-nine contains 3 hundreds,
  6 tens, and 9 units - - - - - - - - - - 369
Nine hundred and ninety-nine contains 9 hundreds,
  9 tens, and 9 units - - - - - - - - - - 999

Nine hundred and ninety-nine is the largest number which is expressed by three figures.

Q. How is one hundred expressed? What place does 1 occupy in this number? How much greater is it in the hundreds' place than in the tens' place? Than in the units' place? How many hundreds, tens, and units in 157? In 369? In 846? In 999? What is the largest number which is expressed by three figures?

8. By methods precisely similar we may express *one thousand, two thousand,* &c., by placing three ciphers on the right of 1, 2, 3, &c. Thus, 1000, 2000, 3000, &c., and all intermediate numbers may be written by writing down in the thousands', hundreds', tens', and units' places, the number of thousands, hundreds, &c., of which they are composed.

We conclude, that any number, whatever be its magnitude, may be expressed by means of the ten characters which have been used ; and this is done by giving different values to these characters, according to the positions which they occupy. Thus,

5 in the units' place, expresses 5 units.

50 is 5 tens, since the 5 is in the tens' place, or 50 units.

500 is 5 hundreds, or 50 tens, or 500 units.

5000 is 5 thousands, or 50 hundreds, or 500 tens, or 5000 units.

The addition of a cipher having the effect in each case to increase the number tenfold.

We may thus form a table, showing the values of the same numbers according to their position, so as to embrace any numbers whatever. Such a table is called a *Numeration Table.* It teaches the manner of reading figures.

Q. How may thousands be expressed by figures? Intermediate numbers? May any number be expressed by the ten known figures? How? How does the addition of a cipher to the right of a number affect it? What does 5 express? 50? 500? 5000? What is a numeration table? What does it teach?

9. NUMERATION TABLE.

Hundreds of Quadrillions.	Tens of Quadrillions.	Quadrillions.	Hundreds of Trillions.	Tens of Trillions.	Trillions.	Hundreds of Billions.	Tens of Billions.	Billions.	Hundreds of Millions.	Tens of Millions.	Millions.	Hundreds of Thousands.	Tens of Thousands.	Thousands.	Hundreds.	Tens.	Units.
																	5.
																6	4.
															3	4	3.
														5,	6	7	2.
													3	8,	6	4	3.
												4	5	3,	2	1	0.
											7,	3	8	9,	0	4	9.
									3	6,	2	0	5,	4	0	2.	
									0	2	4,	5	9	8,	3	9	2.
								8,	2	0	8,	0	8	4,	2	1	0.
							0	5,	3	7	0,	6	5	0,	9	4	1.
						1	3	0,	6	1	4,	2	7	6,	4	3	2.
					7,	3	2	4,	8	5	3,	0	4	7,	8	9	5.
				3	2,	0	1	1,	4	0	5,	9	8	3,	0	2	1.
			5	2	0,	4	7	5,	3	2	0,	1	0	8,	5	0	2.
		7,	9	1	9,	9	3	2,	7	5	3,	2	1	0,	8	7	5.
	2	3,	7	9	1,	8	5	9,	9	2	1,	5	0	9,	3	2	1.
9	4	4,	6	0	8,	7	8	5,	2	6	4,	3	2	1,	8	4	0.

To find the value of any number in this table, we have to see under what denomination the figures which compose it are placed.

Thus, 5 is written under *units*, and is read 5 units.

In the number 64, the 6 is under *tens*, and the 4 under *units;* the number is then 6 tens and 4 units, or sixty-four.

The number 453,210 has the 4 under *hundreds of thousands*, which is 4 hundred thousand; the 5 under *tens of thousands*, which is 50 thousand; the 3 under *thousands*, which is 3 thousand; the 2 under *hundreds*, which is 2 hundred; the 1 under *tens*, which is 1 ten; the 0 under *units*, which is 0 units; the whole number is therefore four hundred and fifty-three thousand, two hundred and ten.

By examining this table, it will be seen, that a number containing but one figure can contain *units* only; two figures, tens and units; three figures, hundreds, tens, and units, &c.

To read any number, we commence on the right and say, units, tens, hundreds, thousands, tens of thousands, hundreds of thousands, millions, &c., until we have passed over all the figures in the number. Thus,

5	3	9	8	6	4
hundreds of thousands.	tens of thousands.	thousands.	hundreds.	tens.	units.

and taking all their values together, we read, five hundred and thirty-nine thousand, eight hundred and sixty-four.

Q. Repeat the numeration table. How is the value of any number determined? If the number 6 fall under thousands, what is its value? If under tens of thousands? Millions? A number with but one figure contains what? Two figures? Three? How do you read any number? What is the figure on the right? The next? Read 124, 764. 2,004,761. 75,030,147. 109,026,374.

EXAMPLES.

I. Write in words upon your slate the *population* of the *six New England States*, viz:

1. *Maine,* - - - 501,793,	4. *Massachusetts,* 737,699,
2. *New Hampshire,* 284,574,	5. *Rhode Island,* 108,830,
3. *Vermont,* - - 291,948,	6. *Connecticut,* - 309,978.

Q. How many New England States? What are they? Which is the largest in population? Which the smallest? What is the population of Massachusetts in round numbers? Rhode Island, &c.? Is your state one of the New England States? What is its population?

II. Write in figures the population of the *four Middle States*, viz:

1. *New York.*—Two millions, four hundred and twenty-eight thousand, nine hundred and twenty-one.

2. *New Jersey.*—Three hundred and seventy-three thousand, three hundred and six.

3. *Pennsylvania.*—One million, seven hundred and twenty-four thousand and thirty-three.

4. *Delaware.*—Seventy-eight thousand and eighty-five.

Q. How many Middle States? What are they? Which is the largest in population? The smallest? The population of New York in round numbers? Delaware? Is your state one of the Middle States? What is its population?

III. Read the population of the *eight Southern States*, viz:

1. *Maryland,*	-	470,019,	5. *Georgia,* - -	691,392,
2. *Virginia,* -	-	1,239,797,	6. *Alabama,* - -	590,756,
3. *North Carolina,*		753,419,	7. *Mississippi,* -	375,651,
4. *South Carolina,*		594,398,	8. *Louisiana,* -	352,411.

Q. How many Southern States? What are they? Which is the largest in population? Which the smallest? Has North Carolina a greater or less number of inhabitants than South Carolina? Is your state one of the Southern States? What is its population?

IV. Write upon your slate the population of the *eight Western States*, in *figures*, viz:

1. *Arkansas.*—Ninety-seven thousand, five hundred and seventy-four.

2. *Tennessee.*—Eight hundred and twenty-nine thousand, two hundred and ten.

3. *Kentucky.*—Seven hundred and seventy-nine thousand, eight hundred and twenty-eight.

4. *Ohio.*—One million, five hundred and nineteen thousand, four hundred and sixty-seven.

5. *Michigan.*—Two hundred and twelve thousand, two hundred and sixty-seven.

6. *Indiana.*—Six hundred and eighty-five thousand, eight hundred and sixty-six.

7. *Illinois.*—Four hundred and seventy-six thousand, one hundred and eighty-three.

8. *Missouri.*—Three hundred and eighty-three thousand, seven hundred and two.

Q. How many Western States? What are they? Which is the largest in population? The smallest? How many states altogether? Which is the largest in population of the twenty-six? Next? Which is the smallest? Is your state one of the Western States? What is its population?

V. Write in figures the population of the *United States* upon your slate, viz: Seventeen millions, sixty-three thousand, three hundred and fifty-three. *Ans.* 17,063,353.

OF THE ROMAN NUMERATION.

10. The above is the most common system of numeration, being used throughout the civilized portion of the world. It is called the *Arabic* method of numeration, as it was invented by the Arabs.

The Romans invented another system, which is sometimes used. They expressed numbers by the seven capitals of the alphabet, viz:

I.	which stands for		1		C.	which stands for			100
V.	"	"	"	5	D.	"	"	"	500
X.	"	"	"	10	M.	"	"	"	1000
L.	"	"	"	50					

The other numbers were expressed by various combinations of these letters, after the following manner:

I.	is	One.	LX.	is	Sixty.
II.	"	Two.	XC.	"	Ninety.
III.	"	Three.	C.	"	One hundred.
IV.	--	Four.	CC.	"	Two hundred.
V.		Five.	D.	"	Five hundred.
VI.		Six.	DC.	"	Six hundred.
VII.	"	Seven.	M.	"	One thousand.
VIII.	"	Eight.	MM.	"	Two thousand.
IX.		Nine.	$\overline{V}$.	"	Five thousand.
X.		Ten.	$\overline{VI}$.	"	Six thousand.
XI.	"	Eleven.	$\overline{X}$.	"	Ten thousand.
XX.	"	Twenty.	$\overline{L}$.	"	Fifty thousand.
XL.	--	Forty.	$\overline{C}$.	"	One hundred thousand.
L.	"	Fifty.	$\overline{M}$.	"	One million.

As often as any character is repeated, so many times is its value repeated. Thus II. is two, XX. is 2 tens, or ~enty, CC. is two hundred, &c.

A less character *before* a greater diminishes its value as much as is denoted by the less. Thus, IV. is one from *five*, or four.

A less character *after* a greater increases its value as much as is denoted by the less. Thus, VI. is five and one, or *six*.

A *bar* ⸺ over a character increases it one thousand fold. V. is five, $\overline{V}$. is five thousand.

Q. What is the common method of numeration called? By whom invented? How extensively is it used? Is there any other method? What called? By whom invented? In what does it consist? How many letters are used? What do they express? How are the other numbers expressed? Write the various numbers from one to twenty upon your slate. From 20 to 50. From 50 to 1000. What effect is produced by repeating a character? Suppose a less character come before a greater? After a greater? What effect has a bar over a character?

ADDITION OF SIMPLE NUMBERS.

11. For the purposes of Arithmetic, numbers are divided into two kinds, viz: *simple* and *compound*.

Simple numbers are numbers which are composed of the same unit, whether the unit be dollars, pounds, yards, or miles. Thus,

4 dollars, 8 dollars, and 10 dollars, is a simple number.

15 pounds, 30 pounds, and 54 pounds, is a simple number.

When the kind of unit is not expressed, the numbers are always simple numbers. Thus, 2, 3, 5, are simple numbers, and they may express so many pounds, or dollars, or other units.

Compound numbers are numbers which are composed of different units. Thus,

4 dollars, 5 cents, and two mills, is a compound number.

6 yards, 2 feet, and 3 inches, is a compound number.

Q. How are numbers divided in arithmetic? What are they? What are simple numbers? Compound? When the kind of unit is not expressed, what kind of numbers are denoted?

12. *Addition, Subtraction, Multiplication,* and *Division,* are the four *fundamental* or *ground* rules of arithmetic. Every question in arithmetic depends upon one or the other of these rules.

Addition consists in expressing by a single number the total value of several numbers.

Addition of simple numbers consists in expressing by a single number the total value of several simple numbers.

Thus, if John has two marbles, and James 3 marbles, they will have between them 5 marbles. Here, by a single number 5 we have expressed the number of marbles which John and James had between them.

Again—if Mary has 3 pins, Jane 4 pins, and Anna 5 pins, they will have altogether 12 pins.

The number which thus expresses the total value of several other numbers is called the *sum.* Thus 5 marbles is the *sum* of 2 marbles and 3 marbles; and 12 pins, the *sum* of 3 pins, 4 pins, and 5 pins.

Q. What are the four ground rules of arithmetic? Why are they important? In what does Addition consist? In what does addition of simple numbers consist? How do you explain this? What is the number called which expresses the total value of several numbers? What is the sum of 2 and 2? 2 and 3? 2 and 4? 2 and 5? 3 and 4? 3 and 5? 3 and 6? 4 and 5? 4 and 7? 4 and 9? 5 and 9?

13. There is no difficulty in adding numbers composed of a single figure. Before proceeding to the addition of larger numbers, every pupil should be able to add promptly those which contain but one figure.

To indicate that two numbers are to be added together, we use the sign $+$, which is called *plus.* Thus $2+3$, denotes that 2 and 3 are to be added together, and is read 2 *plus* 3.

To indicate that two numbers are equal to each other we use the sign $=$, which is called the sign of equality. Thus $2+3=5$, denotes that 2 and 3 added together are *equal* to 5, and is read, 2 plus 3, equal to 5.

What is the sum of the following numbers:

$0 + 1$
$0 + 1 + 2$
$0 + 1 + 2 + 3$
$1 + 2 + 3 + 4$
$1 + 2 + 3 + 4 + 5$
$1 + 2 + 3 + 4 + 5 + 6$
$1 + 2 + 3 + 4 + 5 + 6 + 7$
$1 + 2 + 3 + 4 + 5 + 6 + 7 + 8$
$1 + 2 + 3 + 4 + 5 + 6 + 7 + 8 + 9.$

Find the sum of the above numbers, adding from the right.

Q. What sign is used to indicate that two numbers are to be added together? What is it called? What does it denote? How would you express that 2 and 3 are to be added together? How do you read 2+3? What is the sign of equality? How used? What does it signify? Give an example. What is 2+3 equal to? 2+4? 2+9? 3+4? 3+7? 3+9? 4+3? 4+5? 5+7? 5+9? 8+8? 9+9?

14. Let it now be required to add the numbers 8423 and 1234.

Place the numbers, one under the other, so that units shall be under units, tens under tens, &c. Commencing on the right, we say 4 and 3 are 7, 3 and 2 are 5, 2 and 4 are 6, 1 and 8 are nine, writing down each separate sum, 7, 5, 6, 9, under the numbers which produce it; that is, since 4 units and 3 units make 7 units, 7 must be in the units' place, 3 tens and 2 tens make 5 tens, 5 must be in the tens' place, &c. Hence 9657 is the sum required.

$$\begin{array}{r} 8423 \\ 1234 \\ \hline 9657 \end{array}$$

15. Add the numbers 4897, 821, 1634, together.

Having set down the numbers as before, we say, 4 and 1 are 5, and 7 are 12; now, since 12 units contain 1 ten and 2 units, we only set down the 2 units and carry the 1 ten into the column of tens, and say, 1 we carry, and 3 are 4, and 2 are 6, and 9 are 15; but 15 tens contain 1 hundred and 5 tens, we set down 5 in the column of tens and carry 1 to the column of hundreds, and say, 1 we carry and 6 are 7, and 8 are 15, and 8 are 23, and setting down the 3 hundreds, we carry 2

$$\begin{array}{r} 4897 \\ 821 \\ 1634 \\ \hline 7352 \end{array} = \text{sum.}$$

thousands into the next column and proceed, 2 and 1 are 3, and 4 are 7.   Hence 7352 is the required sum.

16. Add 3909, 1094, 302, and 20, together.

These numbers being set down as before, we have 0 and 2 are 2, and 4 are 6, and 9 are 15, set down 5 and carry 1 ; 1 we carry and 2 are 3, and 0 are 3, and 9 are 12, and 0 are 12, set down 2 and carry 1 ; 1 and 3 are 4, and 0 are 4, and 9 are 13, set down 3 and carry 1 ; 1 and 1 are 2, and 3 are 5. 5325 is the required sum.

$$\begin{array}{r} 3909 \\ 1094 \\ 302 \\ 20 \\ \hline 5325 \end{array} = \text{sum.}$$

17. Add 987641 and 407180 together.

Here the numbers which are carried in each column are indicated by the small figures placed under the columns ; 1 is carried into the column of hundreds and 1 into the column of tens of thousands.

$$\begin{array}{r} 987641 \\ 407180 \\ ^{1}\quad{}^{1} \\ \hline 1394821 \end{array} = \text{sum.}$$

We see by the above examples that if the sum in any one column does not exceed 9, we set down the whole sum under this column ; but if it exceed 9, we set down the right hand figure alone, and carry the remaining figure or figures into the next column.

From these explanations we deduce the following

RULE.

I. *Set down the numbers, one under the other, so that units fall under units, tens under tens, &c., and draw a line beneath them.*

II. *Beginning at the bottom of the right hand column, add up the figures in each column. If the sum in any one column does not exceed 9, write down the whole sum beneath this column ; if it exceed 9, write down the right-hand figure and carry the remaining figure or figures into the next column.*

III. *Finally, under the last column write the entire sum of this column.*

⋂ How are numbers set down for addition ?   Where do you begin
 ? If the sum of the figures in any column do not exceed 9, what
    n ?   If it exceed 9 ?   If the sum in a column be 19, what is
    and what carried ?   If it be 29 ?   30 ?   50 ?   72 ?   99 ?   100 ?
    ꞏet under the last column ?

PROOF OF ADDITION.

18. Addition is proved in several ways.

1st Proof. Begin the addition at the top of the right-hand column, and add each column downwards.

2d Proof. Draw a line under the upper number, add the remaining numbers together, and then add their sum to the upper number.

3d Proof. Write the numbers down in a different order from that in which they appear in the example, and then add as the rule directs. If the example be worked correctly, it will have the same result as that given by the proof.

Add 3024, 721, and 320 together.

	2d Proof.	*3d Proof.*
3024	3024	721
721	———	320
320	721	3024
———	320	———
4065	———	4065
	1041	
	3024	
	———	
	4065	

In the second proof, the numbers 721 and 320 are first added together, and to their sum, 1041, the first number is added. The total sum, 4065, is the same as shown by the example. By the third proof, the number 721 is placed first, 320 next, and then 3024, and the addition gives the same result.

Q. How many ways of proving addition? What is the first proof? Second proof? Third proof? When is the example worked correctly?

EXAMPLES.

1. Add 3094, 5211, 6010, and 4372 together. *Ans.* 18687.

2. Add 1001, 298, 9000, and 32 together. *Ans.* 10331.

3. Add 11040, 3279, 907, 85, and 3 together. *Ans.* 15314.

4. Add 3904, 27520, and 1089753 together. *Ans.* 1121177.

APPLICATIONS.

1. There are 12 calendar months in the year. January has 31 days, February 28, March 31, April 30, May 31, June 30, July 31, August 31, September 30, October 31, November 30, December 31. How many days in the year?

Ans. 365 days.

Q. How many months in the year? What are they? How many days in each? In the year?

2. How many inhabitants were there in the 15 States by the first census, taken in 1790; and what was the total population of the United States, the number of free persons and slaves in each State being as follows:—

States.	No. of Free Persons.	No. of Slaves.	Total in each State.
1. Virginia	454,881 +	293,427 =	*Ans.*
2. Massachusetts	475,257 +	000,000 =	*Ans.*
3. Pennsylvania	430,636 +	3,737 =	*Ans.*
4. North Carolina	293,179 +	100,572 =	*Ans.*
5. New York	318,796 +	21,324 =	*Ans.*
6. Maryland	216,692 +	103,036 =	*Ans.*
7. South Carolina	141,979 +	107,094 =	*Ans.*
8. Connecticut	235,382 +	2,759 =	*Ans.*
9. New Jersey	172,716 +	11,423 =	*Ans.*
10. New Hampshire	141,741 +	158 =	*Ans.*
11. Vermont	85,399 +	17 =	*Ans.*
12. Georgia	53,284 +	29,264 =	*Ans.*
13. Kentucky	61,247 +	11,830 =	*Ans.*
14. Rhode Island	68,158 +	952 =	*Ans.*
15. Delaware	50,209 +	8,887 =	*Ans.*

3,199,556 + 694,480 = 3,894,036,
Total in the United States.

If the total population in each State be correctly determined, the sum of these results will give the total population in the United States.

Q When was the first census taken? How many States at the time? Which was the largest in population? Which the smallest? Which had the most slaves? Next? Which the least? What was the population of the United States in 1790? How many free? Slaves?

3. *The United* States have had *ten* Presidents:—1. General Washington, of Virginia, who served 8 years; 2. John

Adams, of Massachusetts, who served 4 years; 3. Thomas Jefferson, of Virginia, 8 years; 4. James Madison, of Virginia, 8 years; 5. James Monroe, of Virginia, 8 years; 6. John Quincy Adams, of Massachusetts, 4 years; 7. Andrew Jackson, of Tennessee, 8 years; 8. Martin Van Buren, of New York, 4 years; 9. William Henry Harrison, of Ohio, who served 1 month, and died in office; and, 10. John Tyler, of Virginia, who succeeded General Harrison, being at the time Vice-President. How many years since the establishment of the United States government to the commencement of General Harrison's term? *Ans.* 52 years.

Q. How many Presidents have the United States had? Who were they? To what States did they belong? How many years did each serve? Who succeeded Van Buren? Did he live out his term? Who succeeded him on his death? What office did Tyler hold before he became President?

4. The public lands of the United States have been obtained by purchase and cession. What is the total amount that has been purchased and ceded?—the number of acres purchased, and the number ceded, being as follows:—

Ceded by Virginia, New York, Massachusetts, and Connecticut	169,609,819
Ceded by Georgia	58,898,522
Ceded by North and South Carolina......	26,432,000
Purchased of France and Spain	987,852,332

Ans. ‾‾‾‾‾‾‾

Q. How were the public lands of the United States obtained? What States ceded lands? From whom were some of the public lands purchased?

5. What is the total population in the five largest cities in the United States, the population in each being as follows:—New York, 312,710; Philadelphia, 258,037; Baltimore, 102,313; New Orleans, 102,193; Boston, 93,383?

Ans. 868,636.

Q. Which are the five most populous cities in the United States? Which is the largest city? How many inhabitants in each, in round numbers?

6. What was the total expense of the United States government for the year 1842, that of each department being as follows:—*Civil list,* 6,865,451 dollars; *Military establish-*

ment, 8,248,917 dollars; Naval establishment, 7,963,677 dollars?　　　　　　　　　　　*Ans.* 23,078,045 dollars.

Q. Between what departments is the expense of the general government divided? How much is expended in each department? Total expenditure in 1842?

7. What was the total amount of deposits of gold in the United States mints in 1842, it being derived from the following sources:—Virginia mines, 42,163 dollars; North Carolina, 61,629 dollars; South Carolina, 223 dollars; Georgia, 150,276 dollars; Alabama, 5,579 dollars?
　　　　　　　　　　　　　　Ans. 259,870 dollars.

Q. What was the amount of gold deposited in the United States mints in 1842? From what sources derived? Which State yielded the largest amount?

8. What is the total number of inhabitants in the British Provinces in America, the population in each Province being as follows:—*Lower Canada,* 499,739; *Upper Canada,* 506,055; *New Brunswick,* 130,000; *Nova Scotia,* 199,870; *Prince Edward's Island,* 34,666; *Newfoundland,* 81,517; *Honduras,* 3,958?　　　　　　　*Ans.* 1,455,805.

Q. How many British Provinces in America? What are they? The most populous? &c.

SUBTRACTION OF SIMPLE NUMBERS.

19. SUBTRACTION *teaches the method of finding how much one number exceeds another.* This excess is called their *difference,* or the *remainder.* The smaller number is called the *subtrahend,* the larger the *minuend.*

In subtraction of *simple numbers,* the numbers which are considered in any example are supposed to be expressed in terms of the same kind of unit. That is, they are all dollars, or all cents, or all mills.

Q. What does subtraction teach? What is the excess called? Which number is the subtrahend? Which the minuend? What is subtraction of simple numbers?

20. There is no difficulty in the subtraction of numbers which contain but one figure. If John has 6 apples and James 4, any one will promptly tell how many more apples John has than James. He has 2 more apples. If Mary has 9 nuts and Ann 6, Mary has 3 more nuts than Ann.

In the first example, 6 is the *minuend*, 4 the *subtrahend*, and 2 the *remainder;* in the last, 9 is the minuend, 6 the subtrahend, and 3 the remainder.

We can also easily subtract a number containing but one figure from another which is less than 20. Thus, 2 from 11 leaves 9, since 2 and 9 make 11 ; 6 from 13 leaves 7, since 6 and 7 make 13 ; 8 from 17 leaves 9, since 8 and 9 make 17.

To indicate the subtraction of two numbers, we use the sign —, which is called *minus*. Thus 6—4 denotes that 4 is to be taken from 6, and is read 6 *minus* 4.

Q. 2 from 4 leaves how many ? 2 from 5 ? 2 from 6 ? 2 from 8 ? 2 from 10 ? 3 from 5 ? 3 from 10 ? 3 from 15 ? Which is the minuend in these examples ? The subtrahend ? The remainder ? What sign is used to indicate subtraction ? What is it called ? How do you read 7—3 ? What is 6—4 equal to ? 7—5 ? 8—3 ? 12—8 ? 16—4 ? 18—9 ? 19—7 ?

21. Let it be required to subtract 3456 from 6598. Place the numbers under each other, as in addition, units under units, tens under tens, &c. We commence on the right, and say, 6 from 8 leaves 2 ; 5 from 9 leaves 4 ; 4 from 5 leaves 1 ; 3 from 6 leaves 3 ; setting each separate remainder, 2, 4, 1, 3, under the numbers which produce it. Thus, 6

6598 minuend,
3456 subtrahend,
3142 remainder.

units taken from 8 units leaves 2 units ; the 2 is set in the *units* place : 5 tens from 9 tens leaves 4 tens ; 4 is set in the *tens* place ; and so on. 3142 is the remainder.

22. Take 7652 from 9324. 2 from 4 leaves 2 ; but we cannot take 5 from 2, since it is greater than 2. To avoid this difficulty, we borrow 1 from the *hundreds* of the *minuend*, and

9324
7652
1672 remainder.

add it to the 2 tens ; but 1 hundred is the same as 10 tens ; hence, adding 10 to 2, we have 12, from which if 5 be taken, the remainder will be 7. But as we borrowed 1 from the hundreds of the *minuend*, the value of the remainder will not be changed, if we add 1 to the hundreds of the subtrahend : 1 and 6 are 7 ; 7 from 3 we cannot, but borrowing 1 from the thousands, or adding 10 to the 3 hundreds, we have 13 :

B

7 from 13 leaves 6 ; 1 we carry to 7 makes 8 ; 8 from 9 leaves 1 : the remainder required is 1672.

23. From these examples we see that when the units, tens, hundreds, &c., of the minuend are greater than those of the subtrahend, the subtraction may be immediately performed by taking the less figure from the greater. But when a figure in the subtrahend is greater than the one corresponding to it in the minuend, we add 10 to the figure in the minuend, and from this sum take the figure in the subtrahend ; then carry 1 to the next figure in the subtrahend, before it is taken from the figure above it.

The following examples will illustrate these principles. The small 10 above the top lines shows where 10 has been added, and the 1 below the figures in the subtrahend shows where 1 has been carried.

$$
\begin{array}{r}
\overset{10\ 10\ 10\ 10}{8\,2\,4\,6\,7} \\
2\,9\,6\,8 \\
\underset{1\ 1\ 1\ 1}{} \\
\hline
7\,9\,4\,9\,9 \text{ remainder.}
\end{array}
\qquad
\begin{array}{r}
\overset{10\ 10\quad 10\ 10}{2\,3\,4\,6\,8\,4} \\
1\,8\,6\,2\,8\,6 \\
\underset{1\ 1\ \ 1\ 1}{} \\
\hline
0\,4\,8\,3\,9\,8 \text{ remainder.}
\end{array}
$$

In the first example we say, 8 from 17 leaves 9 ; 1 carried to 6 makes 7 ; 7 from 16 leaves 9 ; 1 carried to 9 makes 10 ; 10 from 14 leaves 4 ; 1 carried to 2 makes 3 ; 3 from 12 leaves 9 ; 1 we carry from 8 leaves 7. In like manner with the second example. We have therefore the following

RULE.

I. *Place the number to be subtracted under the number from which it is to be taken, as in addition, and draw a line beneath them.*

II. *Beginning on the right, subtract each figure from the one just over it, and set down the remainder.*

III. *When any figure in the subtrahend is greater than the one over it in the minuend, add 10 to the less figure, and then subtract, carrying 1 to the next figure in the subtrahend.*

Q. How are the numbers set down for subtraction ? Where do you begin to subtract ? How do you subtract ? If a figure in the minuend is less than the one below it, what do you do ?

24. *Proof.*—Add the remainder to the subtrahend; if the work is correct, the sum will be equal to the minuend.

Q. How is subtraction proved? When is the work correct?

EXAMPLES.

1. From 1000 take 12.   *Ans.* 988.
2. From 20000 take 14.   *Ans.* 19986.
3. From 7002 take 3495.   *Ans.* 3507.
4. From 11 thousand 11 hundred and 11 take 11.   *Ans.* 12100.
5. From 11011 take 99.   *Ans.* 10912.
6. Take 10 from 10 million.   *Ans.* 9999990.

APPLICATIONS.

1. America was discovered by Christopher Columbus in the year 1492, and the Constitution of the United States adopted by the States in 1789. How many years between these two events?   *Ans.* 297 years.

Q. By whom was America discovered? In what year? When was the Constitution of the United States adopted?

2. The first settlement in the United States was made by the English at *Jamestown*, Virginia, in the year 1607; General Washington died in 1799. What period between these two events?   *Ans.* 192 years.

Q. By whom was the first settlement in the United States made? Where made? In what year? When did General Washington die?

3. Our Revolutionary war commenced in 1775, and ended in 1783; the last war with Great Britain commenced in 1812 and ended in 1815. How much longer was the Revolutionary war than the last war?   *Ans.* 5 years.

Q. When did the Revolutionary war commence? When did it end? When did the last war with Great Britain commence? When end? How long did each war last?

4. The mariners' compass was discovered in Italy, in 1302; how many years before the declaration of American independence, in 1776?   *Ans.* 474 years.

Q. When was the mariners' compass discovered? Where? When was the declaration of American independence published?

5. The revenue of the Post Office in 1842 was 4,546,246 dollars, and its expenditure 4,235,052 dollars. How much did the revenue exceed the expense? *Ans.* 311,194 dollars.

Q. Was the Post Office department a source of expense or revenue in 1842?

6. The population of the United States in 1840 wa 17,063,353, of which 2,487,355 were slaves. How man free persons in the United States? *Ans.* 14,575,998.

Q. State the total population of the United States in round numbers. How many of these are slaves?

7. What has been the number of pounds of coffee left for home consumption in the United States annually, the amount imported and exported each year from 1821 to 1838, being as follows;—

	Years.	Imports.* lbs.	Exports. lbs.	Home Consumption. lbs.
1.	1821.	21,273,659—	9,387,596=	*Ans.* 11,886,063
2.	1822.	25,782,390—	7,267,119=	*Ans.* 18,515,271
3.	1823.	37,337,732—	20,900,687=	*Ans.*
4.	1824.	39,224,251—	19,427,227=	*Ans.*
5.	1825.	45,190,630—	24,512,568=	*Ans.*
6.	1826.	43,319,497—	11,584,713=	*Ans*
7.	1827.	50,051,986—	21,697,789=	*Ans.*
8.	1828.	55,194,697—	16,037,964=	*Ans.*
9.	1829.	51,133,538—	18,083,843=	*Ans.*
10.	1830.	51,488,248—	13,124,561=	*Ans.*
11.	1831.	81,759,386—	6,056,629=	*Ans.*
12.	1832.	91,729,329—	55,251,158=	*Ans.*
13.	1833.	99,955,020—	24,897,114=	*Ans.*
14.	1834.	80,153,366—	35,806,861=	*Ans.*
15.	1835.	103,199,572—	11,446,775=	*Ans.*
16.	1836.	93,790,506—	16,143,207=	*Ans.*
17.	1837.	88,140,403—	12,096,332=	*Ans.*
18.	1838.	88,139,720—	5,267,087=	*Ans.*

Q. Have the United States imported more coffee than they have consumed? Have the imports increased or diminished? Have the exports been uniform? Has the home consumption increased or diminished?

---

* *Imports* are articles brought *into* a country. *Exports* are articles shipped *from* a country.

APPLICATIONS IN ADDITION AND SUBTRACTION.

1. Find how much the value of the *imports* of the following States, in 1839, exceeded their *exports,* and what was the *total* excess of the imports over the exports.

State, &c.	Value of Imports. Dollars.	Value of Exports. Dollars.	Excess. Dollars.
Maine.....	982,724—	895,485=*Ans.*	87,239
Vermont...	413,513—	193,886=*Ans.*	219,627
Mass......	19,385,223—	9,276,085=*Ans.*	
R. Island..	612,057—	185,234=*Ans.*	
New York.	99,882,438—	33,268,099=*Ans.*	
Pennsylv'a.	15,050,715—	5,299,415=*Ans.*	
Maryland..	6,995,285—	4,576,561=*Ans.*	
Kentucky..	10,480—	3,723=*Ans.*	
Michigan..	176,221—	133,305=*Ans.*	
Total..	143,508,656—	53,831,793=*Ans.*	89,676,863

2. Find how much the value of the *exports* of the following States and Territories, in 1839, exceeded their *imports,* and what was the total excess of the exports over the imports.

State.	Value of Exports. Dollars.	Value of Imports. Dollars.	Excess. Dollars.
N. Hamp..	81,944—	51,407=*Ans.*	30,537
Connecticut	583,226—	446,191=*Ans.*	137,035
N. Jersey..	98,079—	4,182=*Ans.*	
Dist. Colum.	503,717—	132,511=*Ans.*	
Virginia...	5,187,196—	913,462=*Ans.*	
N. Carolina	427,926—	229,233=*Ans.*	
S. Carolina.	10,385,426—	3,086,077=*Ans.*	
Georgia...	5,970,443—	413,987=*Ans.*	
Alabama..	10,338,159—	895,201=*Ans.*	
Louisiana..	33,181,167—	12,064,942=*Ans.*	
Ohio......	95,854—	19,280=*Ans.*	
Florida....	334,806—	279,893=*Ans.*	
Total...	67,187,943—	18,536,366=*Ans.*	48,651,577

Q. Which is the largest importing State? The smallest? Do the New England States import or export most? How is it in the South? The Middle States? The Western? Are the exports of your State greater or less than *its* imports? Which are the principal exporting States? Importing States?

3. What is the supposed total population of the globe, and its excess over that of America; the population of the *four quarters* of the globe being as follows:—*Europe*, 283,240,043; *Asia*, 608,516,019; *Africa*, 101,498,411; *America*, 48,007,150?

Ans. Total population,   991,261,623.
Excess over America, 943,254,473.

Q. What are the four quarters of the globe? Which is the most opulous? The least? How many inhabitants in the world?

4. The population of France is computed at 33,600,000 inhabitants; how much does it exceed that of Great Britain, which is composed as follows :—England and Wales, 13,897,187; Scotland, 2,365,114; Ireland, 7,767,401?

Ans. 9,570,298.

Q. What countries compose the kingdom of Great Britain? Which is the most populous? The least? Has France more inhabitants than Great Britain?

5. The debt of Great Britain in 1837 was 787,638,916 pounds sterling, of which amount 121,267,993 were contracted during our Revolutionary war, and 601,500,343 during the war with the French which commenced in 1793. What would have been the debt of Great Britain, had not these wars taken place? *Ans.* 64,870,580 pounds sterling.

Q. How was the greater portion of the debt of Great Britain contracted? What was the expense of the American war to her? Of the French war of 1793?

6. Of the public lands of the United States there have been sold 107,796,536 acres; 33,756,559 acres have been granted for *internal improvements, education, military services,* and *reservations.* There were originally 1,242,792,673 acres. How many acres yet unsold?

Ans. 1,101,239,578 acres.

Q. How have the public lands been disposed of? How many acres sold? How many acres granted? For what purposes? How many acres unsold?

7. The total amount of the sales of the public lands to 1843 was 170,940,942 dollars, of which sum 68,524,991 dollars were paid for the Indian title, and for the purchase of Florida from Spain and Louisiana from France; and *9,966,610 dollars* were paid for making the surveys and

selling the lands. How much has been paid into the Treasury of the United States? *Ans.* 92,449,341 dollars.

Q. From whom were public lands purchased? How was Florida obtained? Louisiana? What has the United States government paid in this way? What is done with the nett proceeds of the public lands? How much has been paid into the United States Treasury?

8. Find the total number of inhabitants in the United States at each of the six enumerations, the increase in each period of ten years, and the total increase since 1790, the population being as stated below:—

States.	1790.	1800.	1810.	1820.	1830.	1840.
Maine............	96,540	151,719	228,705	298,335	399,955	501,793
New Hampshire...	141,899	183,762	214,360	244,161	269,328	284,574
Vermont.........	85,416	154,465	217,713	235,764	280,652	291,948
Massachusetts.....	378,717	423,245	472,040	523,287	610,408	737,699
Rhode Island......	69,110	69,122	77,031	83,059	97,199	108,830
Connecticut.......	238,141	251,002	262,042	275,202	297,665	309,978
New York........	340,120	586,756	959,949	1,372,812	1,918,608	2,428,921
New Jersey........	184,139	211,949	249,555	277,575	320,823	373,306
Pennsylvania.....	434,373	602,365	810,091	1,049,458	1,348,233	1,724,033
Delaware.........	59,098	64,273	72,674	72,749	76,748	78,085
Maryland........	319,728	341,548	380,546	407,350	447,040	470,019
Virginia.........	748,308	880,200	974,622	1,065,379	1,211,405	1,239,797
North Carolina....	393,751	478,103	555,500	638,829	737,987	753,419
South Carolina....	249,073	345,591	415,115	502,741	581,185	594,398
Georgia..........	82,548	162,101	252,433	340,987	516,823	691,392
Alabama.........	· ·	· ·	·	20,845	127,901	309,527	590,756
Mississippi.......	· ·	8,850	40,352	75,448	136,621	375,651
Louisiana........	· ·	· ·	76,556	153,407	215,739	352,411
Arkansas..........	· ·	· ·	·	14,273	30,388	97,574
Tennessee........	30,791	105,602	261,797	422,813	681,904	829,210
Kentucky........	73,077	220,955	406,511	564,317	687,917	779,828
Ohio............	· ·	45,365	230,760	581,434	937,903	1,519,467
Michigan.........	· ·	· ·	4,762	8,896	31,639	212,267
Indiana..........	· ·	4,875	24,520	147,178	343,031	685,866
Illinois..........	· ·	·	12,282	55,211	157,445	476,183
Missouri.........	· ·	·	20,845	66,586	140,445	383,702
Dist. of Columbia.	· ·	14,093	24,023	33,039	39,834	43,712
Florida..........	· ·	· ·	·	·	34,730	54,477
Wisconsin........	· ·	· ·	·	·	· ·	30,945
Iowa............	· ·	· ·	·	·	· ·	43,112
	3,924,829	5,305,941				

Ans. First 10 years' increase............. 1,361,112
Second 10 years' increase........... 1,959,618
Third 10 years' increase............ 2,872,682
Fourth 10 years' increase.......... 3,223,001
Fifth 10 years' increase............ 4,202,161

Total increase,........... 13,138,524

9. The revenue of the United States for the year 1842, de

rived from *customs, public lands, Treasury notes, loans,* and miscellaneous sources, was estimated at 34,502,593 dollars, and there was a balance in the Treasury, January, 1842, of 230,483 dollars. The expenditures for the same period were estimated at 35,308,634 dollars. What was the excess of expenditures over the revenue?   *Ans.* 575,558 dollars.

Q. From what sources was the revenue of the United States derived in 1842? Was it equal to the expenditures? What difference?

10. For the year ending July 1, 1842, there were paid 737,605 dollars for the transportation of the United States mail by *horse and sulky,* 1,700,510 dollars by *stage,* 649,581 dollars by *steamboat and railroad,* and 20,000 dollars for *mail bags;* the total expenditure for the same period was 4,235,052. What was the amount paid for other purposes?

Ans.

Q. How is the United States mail transported? To which kind of conveyance is the most money paid?

MULTIPLICATION OF SIMPLE NUMBERS.

25. *To multiply one number by another, is to repeat the first number as many times as there are units in the second.* Thus, to multiply 3 by 4, is to take 3 as many times as there are units in 4, that is, 4 times; or $3+3+3+3=12$.

The number to be multiplied is called the *multiplicand,* the number by which it is multiplied the *multiplier,* and the result the *product.* In the above example, 3 is the multiplicand, 4 the multiplier, and 12 the product.

The multiplicand and multiplier are called the *factors* of the product, because they produce the product.

In multiplication of *simple* numbers, the numbers in any one example are all dollars, or all pounds.

Q. What is meant by multiplying one number by another? To multiply 3 by 4, is equivalent to what? What is the number called that is to be repeated? What is the multiplier? What is the product? In multiplying 3 by 4, which is the multiplier? Which the multiplicand? What is their product? What are the multiplier and multiplicand called? *Why are they called* factors? In multiplication of simple numbers, *what kind of numbers* are considered in any one example?

26. To indicate multiplication we use the sign $\times$. Thus, 3×4 denotes that 3 is to be multiplied by 4.

From what has been said, multiplication consists simply in writing down the multiplicand as many times as there are units in the multiplier, and then adding up all these numbers for the product.

Products.

Thus, 1×1 $\qquad = 1.$

$\qquad 1 \times 2 = 1 + 1 \qquad = 2.$

$\qquad 1 \times 3 = 1 + 1 + 1 \qquad = 3.$

$\qquad 1 \times 4 = 1 + 1 + 1 + 1 \qquad = 4.$

$\qquad 1 \times 5 = 1 + 1 + 1 + 1 + 1 \qquad = 5.$

$\qquad 2 \times 1 \qquad = 2.$

$\qquad 2 \times 2 = 2 + 2 \qquad = 4.$

$\qquad 2 \times 3 = 2 + 2 + 2 \qquad = 6.$

$\qquad 3 \times 4 = 3 + 3 + 3 + 3 \qquad = 12.$

$\qquad 5 \times 6 = 5 + 5 + 5 + 5 + 5 + 5 \qquad = 30.$

$\qquad 9 \times 5 = 9 + 9 + 9 + 9 + 9 \qquad = 45.$

By adding up in this way the various numbers from 1 to 12,—1, 2, 3, 4, 5, 6, 7, 8, 9, 10, 11, 12 times,—we may readily obtain the different products arising from the multiplication of all the numbers from 1 to 12, by the numbers 1, 2, 3, 4, 5, 6, 7, 8, 9, 10, 11, 12.

Q. How is multiplication indicated? To indicate that 3 is to be multiplied by 4, how is the sign used? What does the sign $\times$ denote? Multiplication consists simply in what? How do you ascertain the product of one number by another? In 3×4, how many times would you repeat 3? In 3×12? Might the products of all numbers be obtained in this way? What is the product of 3×10? 5×10? 7×10? 8×9? 8×8? 8×7? 7×8? 12×5? 4×12? 9×9? 9×10? 9×12? 9×13? 10×10? 10×11? 10×12? 11×10? 11×9? 11×12? 12×8? 12×9? 12×10? 12×11? 12×12?

27. As the products of the numbers from 1 to 12 are constantly needed in the multiplication of larger numbers, it has been found convenient to form them into a table, called the *Multiplication Table.* By committing this table carefully to memory, there will be no difficulty in multiplying numbers of any magnitude.

B 2

MULTIPLICATION TABLE.

1	2	3	4	5	6	7	8	9	10	11	12
2	4	6	8	10	12	14	16	18	20	22	24
3	6	9	12	15	18	21	24	27	30	33	36
4	8	12	16	20	24	28	32	36	40	44	48
5	10	15	20	25	30	35	40	45	50	55	60
6	12	18	24	30	36	42	48	54	60	66	72
7	14	21	28	35	42	49	56	63	70	77	84
8	16	24	32	40	48	56	64	72	80	88	96
9	18	27	36	45	54	63	72	81	90	99	108
10	20	30	40	50	60	70	80	90	100	110	120
11	22	33	44	55	66	77	88	99	110	121	132
12	24	36	48	60	72	84	96	108	120	132	144

The *upper* line of this table consists of the numbers from 1 to 12 in order. The *second* line is formed by adding each of the numbers of the first line to itself, and therefore contains each number in the first line *doubled*, that is, the product of each by two.

The *third* line is formed by adding the products in the second line to the numbers in the first line, and the several sums will be the products of the numbers in the first line by 3.

In like manner the other lines are formed; and the last line will represent the products of 1, 2, 3, 4, 5, 6, 7, &c., by 12.

This table is read as follows:—2 times 1 are 2, 2 times 2 are 4, 2 times 3 are 6, 2 times 4 are 8, 2 times 5 are 10, 2 times 6 are 12, 2 times 7 are 14, 2 times 8 are 16, 2 times 9 are 18, 2 times 10 are 20, 2 times 11 are 22, 2 times 12 are 24. Again, 3 times 1 are 3, 3 times 2 are 6, 3 times 3 are 9, &c.: 4 times 1 are 4, 4 times 2 are 8, 4 times 3 are 12, &c.

Q. What is the Multiplication Table? *Why is it used? How is the top line formed? The second line? What do the numbers in the* second line denote? How is the third line formed? What do the numbers in this line denote? Read the Multiplication Table. How much is 6 times 8? 8 times 6? 10 times 11? 4×5? 12×10? 10×12? 11×10? 11×11? 11×12? 12×12?

28. The product of the numbers is the same by taking either of them as the multiplier.

Thus, $3 \times 7 = 3+3+3+3+3+3+3 = 21.$
$\qquad 7 \times 3 = 7+7+7 \qquad\qquad = 21.$

Q. Is the product of two numbers affected by making the multiplicand the multiplier, and the multiplier the multiplicand? Explain this. Will 3×7 be greater or less than 7 times 3? Will it be equal to it? Why? What is the product of 9 by 8? Of 8 by 9? Of 9 by 10? Of 10 by 9?

29. Let it now be required to multiply 8459 by 7.

Writing the multiplicand, 8459, 7 times, as we have done, and adding up the units, tens, &c., we find the result to be 59213, which is the product required. But it is evident that this operation reduces itself to taking 7 times the 9 units, 7 times the 5 tens, 7 times the 4 hundreds, and 7 times the 8 thousands, and taking afterwards the sum of these products. Knowing, from the Multiplication Table, the products of 9, 5, 4, and 8 by 7, we may obtain the product of 8459 by 7 as follows:—

$$
\begin{array}{r}
8459 \\
8459 \\
8459 \\
8459 \\
8459 \\
8459 \\
8459 \\
\hline
59213
\end{array}
$$

Place the multiplier under the multiplicand, and draw a line beneath:— 7 times 9 are 63, that is, 6 tens and 3 units; we set down the 3 in the units' place, and retain the 6 to be carried into the next product; 7 times

$$
\begin{array}{r}
8459 \text{ multiplicand,} \\
7 \text{ multiplier,} \\
\hline
59213 \text{ product.}
\end{array}
$$

5 are 35, and 6 we carry are 41, or 4 hundreds and 1 ten; set down the 1 in the tens' place, and retain the 4 to be carried when we multiply the hundreds; 7 times 4 are 28, and 4 we carry are 32 hundreds, or 3 thousands and 2 hundreds; set down 2 under the hundreds, and retain 3; 7 times 8 are 56, and 3 we carry are 59; here we set down the whole number 59, as there are no more numbers to multiply. The product is therefore 59213, as we before found.

30. It appears from this example that multiplication shortens very much the addition of quantities, and that by means of the Multiplication Table we may determine the product of numbers by much fewer figures than if we were to go through the addition indicated.

Multiplication may then be considered as a short method of performing addition.

We distinguish two cases in multiplication: 1. Where the multiplier does not exceed the limit of the Multiplication Table, which is 12; and, 2. Where it exceeds this limit, or is greater than 12.

Q. What operation is shortened by means of multiplication? What may multiplication be considered? How many cases in multiplication? What is the first case? Second case?

CASE I.

SHORT MULTIPLICATION—THE MULTIPLIER NOT GREATER THAN 12.

31. The example just explained enables us to form the following

RULE.

I. *Set the multiplier under the units' place of the multiplicand.*

II. *Multiply the units of the multiplicand by the multiplier, and if the product contain only units, set them down in the units' place of the product. If the product contain units and tens, set down the units and retain the tens to be carried.*

III. *Multiply the tens of the multiplicand by the multiplier, and add to the product the tens which were retained (if any); set down the tens and retain the hundreds (if any), and thus proceed until every figure in the multiplicand has been multiplied. When the last figure is multiplied, set down the whole result.*

Q. What is the rule for short multiplication? How is the multiplier set down? When do you carry? When not?

EXAMPLES.

1. Multiply 6436 by 4.

Commencing on the right, we say, 4 times 6 are 24, set down 4 and carry 2; 4 times 3 are 12, and 2 we carry are 14, set down 4 and carry 1; *4 times 4 are 16*, and 1 we carry are 17, set

$$\begin{array}{r} 6436 \\ 4 \\ \hline 25744 \end{array}$$

down 7 and carry 1; 4 times 6 are 24, and 1 we carry are 25; the product is 25744.

2. Multiply 704 by 5.

5 times 4 are 20, set down 0 and carry 2; 5 times 0 are 0, and 2 we carry are 2, set down 2; 5 times 7 are 35, which set down; the product is 3520.

$$\begin{array}{r} 704 \\ 5 \\ \hline 3520 \end{array}$$

3. Multiply 5004 by 6.   *Ans.* 30024.

4. Multiply 372186 by 8.   *Ans.*

5. Multiply 9097030 by 9.   *Ans.*

APPLICATIONS IN SHORT MULTIPLICATION.

1. The Congress of the United States is composed of a Senate and House of Representatives. The Senate has 52 members, being 2 from each of the 26 States; and there are 242 members in the House: each member receives 8 dollars per day during the session of Congress. What is the expense of the Senate and House of Representatives, each, per day?   *Ans.* Senate,   416 dollars.
H. of R. 1936 dollars.

Q. How is the Congress of the United States composed? How many members in the Senate? In the House of Representatives? How many senators has each State? What is the pay of a member per day? Is he paid when Congress is not in session?

2. The Legislatures of all of the States are composed of a Senate and House of Representatives elected by the people. What is the daily expense of the Legislatures of the following States?—the number of senators and representatives, and the pay of each per day, being as follows:—

States.	No. of Senators.		No. of Representatives.		Pay. Dollars.	Total Expense Dollars.
New York,	32	+	128	×	3	=*Ans.* 480
Virginia,	32	+	134	×	4	=*Ans.*
Delaware,	9	+	21	×	4	=*Ans.*
Louisiana,	17	+	60	×	4	=*Ans.*
Kentucky,	38	+	100	×	3	=*Ans.*

Q. How are the Legislatures of the States composed? By whom elected? How many senators has New York? Virginia, &c.? How many representatives has New York? Virginia? What is the pay of members per day in New York? Virginia, &c.?

3. Find the total value of the flour shipped from the United States from 1835 to 1838 inclusive; the number of barrels shipped each year, and the prices per barrel, being as follows ;—

	Barrels.	Price.		Value.
1835.	779,396 × 5	dollars=		
1836.	505,400 × 8	dollars=		
1837.	318,719 × 9	dollars=		
1838.	448,161 × 7	dollars=		

Ans. 13,945,778 dollars.

CASE II.

LONG MULTIPLICATION — THE MULTIPLIER GREATER THAN 12.

32. Multiply 347 by 13.

Set down the multiplier under the multiplicand, placing units under units, &c. 3 times 7 are 21, set down 1 and carry 2; 3 times 4 are 12, and 2 we carry are 14, set down 4 and carry 1; 3 times 3 are 9 and 1 are 10. The product of 347 by 3 is therefore 1041. Multiply now the 7 by the 1 ten, the product is 7 tens, set down the 7 in the tens' place; 4 tens multiplied by 1 ten give 4 hundreds, set down the 4 in the hundreds' place; 3 hundreds multiplied by 1 ten will produce 3 thousands, set down the 3 in the thousands' place. The sum of these two products is 4511, which is the product required.

```
 347
  13
————
1041
 347
————
4511
```

33. Multiply 406 by 307.

In this example we say, 7 times 6 are 42, set down 2 and carry 4; 7 times 0 are 0, and 4 we carry are 4, set down 4; 7 times 4 are 28, set down 28. Then beginning with 0, 0 times 6 are 0, set down 0; 0 times 0 are 0, set down 0; 0 times 4 are 0, set down 0. Multiplying now by 3 hundreds, 3 hundreds multiplied by 6 units is 18 hundreds, set down 8 under the hundreds and carry 1; 3 times 0 are 0, and 1 we carry are 1, set down 1; 3 times 4 are 12. The addition of these products *gives 124642* for the product required.

```
   406
   307
—————
  2842
   000
  1218
—————
124642
```

From these examples we see that each successive multiplication by the figures of the multiplier removes the product one place to the left; and that when a 0 appears in the multiplier, we need not multiply by it, but may pass at once to the next figure in the multiplier, observing to set down the right-hand figure of the product under the figure by which we multiply.

Q. How are the numbers of the multiplier set down in Long Multiplication? What is the effect of the successive multiplication by the figures of the multiplier? When a 0 appears in the multiplier, is it necessary to multiply by it? When you multiply by the next figure, how is the product set down?

From these examples we deduce the following

RULE FOR LONG MULTIPLICATION.

34. I. *Write the multiplier under the multiplicand, placing units under units, tens under tens, &c., and draw a line beneath.*

II. *Commencing on the right, multiply each figure of the multiplicand by each figure of the multiplier, setting down the first figure of the product by the units' figure under units, by the tens' figure under tens, &c.*

III. *Add up the partial products; the sum will be the total product required.*

PROOF.—*Perform the operation by using the multiplicand as the multiplier; the product will be the same if the work be correct.*

Q. What is the rule for Long Multiplication? How are the figures set down? Where do you commence to multiply? Where is the first figure of each partial product set down? When you have multiplied by all the figures in the multiplier, how is the total product obtained? How is multiplication proved?

EXAMPLES.

1. Multiply 456 by 39.   *Ans.* 17784.

2. Multiply 111 by 111.   *Ans.* 12321.

3. Multiply 86972 by 1208.   *Ans.* 105062176.

4. Multiply 870497 by 500407.

5. Multiply 207392 by 7000401.

6. Multiply 99864208 by 2259.

7. Multiply 552310764 by 3331.

8. Multiply 999953287 by 59.

9. Multiply 11 thousand 11 hundred and 11 by 4 thousand and 4.   *Ans.* 48492444.

APPLICATIONS IN LONG MULTIPLICATION.

1. What was the value of the tobacco annually exported from the United States from 1821 to 1840, the number of *h*ogsheads, and the average value of each hogshead, for each year, being as follows :—

Years.	Number of Hogsheads.		Average Value per Hogshead. Dollars.		Total Value. Dollars.
1821.	66,858	×	84	=	5,616,072
1822.	83,169	×	74	=	
1823.	99,009	×	63	=	
1824.	77,883	×	62	=	
1825.	75,984	×	80	=	
1826.	64,098	×	83	=	
1827.	100,025	×	65	=	
1828.	96,278	×	54	=	
1829.	77,101	×	64	=	
1830.	83,810	×	66	=	
1831.	86,718	×	56	=	
1832.	106,806	×	56	=	
1833.	83,153	×	69	=	
1834.	87,979	×	74	=	
1835.	94,353	×	87.	=	
1836.	109,442	×	91	=	
1837.	100,232	×	57	=	
1838.	100,593	×	73	=	
1839.	78,995	×	124	=	
1840.	119,484	×	81	=	

Q. Is tobacco one of the exports of the United States? How many hogsheads are exported in each year, on the average? In what years was the largest number exported? In what year was the highest value per hogshead? In what year the lowest?

2. The tobacco exported from the United States is principally raised in *Maryland, Virginia, Kentucky,* and *Ohio;* and is sent chiefly to *England, France, Holland,* and *Germany.* What was the estimated value of the tobacco exported *to these* countries from 1821 to 1840, the number of

hogsheads, and the average value per hogshead, for the 20 years, being as follows:—

Countries.	Hogsheads.	Val. per Hogs'd. Dollars.	Total Value. Dollars.
England........	524,640 ×	95	=49,840,800
France	146,834 ×	111	=
Holland	423,707 ×	51	=
Germany......	573,918 ×	59	=
Other countries .	322,901 ×	74	=

Q. Which of the States produce tobacco? To what countries is it usually imported? To which country is the most tobacco sent?

35. *Remark* 1.—When a number has to be multiplied by 10, 100, 1000, &c., the multiplication may be at once effected by placing on the right of the multiplicand as many ciphers as there are in the multiplier, since annexing one cipher to a number increases it *ten* times, two ciphers one *hundred* times, &c. (*Art.* 8.)

Thus, the product of 655 by 10 is 6550; that of 655 by 100 is 65500, &c.

Q. How may the multiplication be at once effected, when the multiplier is 10? 100? 1000? What is the reason of this? What is the product of 24 by 10? 89 by 10000? 94 by 100000?

36. *Remark* 2.—If one or both the numbers to be multiplied have ciphers on the right hand, we may neglect the ciphers and multiply the significant figures, then place on the right of the product as many ciphers as there are in the multiplicand and multiplier.

Thus multiply 7400 by 2000.

We neglect the ciphers in this example, and multiply 74 by 2, and to the product, 148, annex 5 ciphers, since there are 3 in the multiplier and 2 in the multiplicand.

$$\begin{array}{r} 7400 \\ 2000 \\ \hline 14800000 \end{array}$$

Q. When there are ciphers on the right of the two factors, how may you multiply? How many ciphers do you annex to the product in multiplying 32000 by 30000? 40 by 2000? 84 by 500? 2 by 100000?

37. *Remark* 3.—When the multiplier can be separated into factors, the multiplication may be performed by multiplying by each of these factors successively: the last product will be the total product.

Thus, let it be *required* to multiply 345 by 72.

The number 72 is equal to 8 times 9; 8 and 9 are then

the factors of 72. 12 times 6 also produce 72; 12 and 6 are its factors also; or 72 may be produced by multiplying 12 by 2 and afterwards by 3; and in this case 2, 3 and 12 would be its factors. The product of 345 by 72 may there-fore be obtained by multiplying successively by either of these three sets of factors.

Operation by factors 8 and 9.	*By 12 and 6.*	*By 2, 3, and 12.*
345	345	345
8	12	2
2760	4140	690
9	6	3
24840 product.	24840 product.	2070
		12
		24840 product.

Q. When the multiplier can be separated into factors, how may the multiplication be performed? If 72 were the multiplier, what would be its factors? By what would you multiply? Which would be the total product? What are the factors of 48? Of 24? Of 64? Of 100? Of 144? Of 132?

EXAMPLES UPON THE REMARKS.

1. Multiply 874 by 10000.   *Ans.* 8740000.

2. Multiply 1201 by 1000.   *Ans.* 1201000.

3. Multiply 410200 by 200.   *Ans.* 82040000.

4. Multiply 30700 by 101000.   *Ans.*

5. Multiply 4104 by 48, its factors being 6 and 8, or 12 and 4.

6. Multiply 10010 by 96, its factors being 12 and 8.

APPLICATIONS.

1. The number of members of the United States House of Representatives is fixed by Congress once in 10 years. What would be the apparent population of the United States, at each of the following periods of apportionment, the num-ber of members, and the number of inhabitants to each member, being as follows :—

Years.	Inhabitants to each Member.		Number of Members.		Total Inhabitants.
1793.	33,000	×	105	= Ans.	3,465,000
1803.	33,000	×	141	= Ans.	4,653,000
1813.	35,000	×	181	= Ans.	6,335,000
1823.	40,000	×	212	= Ans.	
1833.	47,700	×	242	= Ans.	
1843.	70,680	×	223	= Ans.	

Q. How many members in the present House of Representatives? How many inhabitants to a member? By whom fixed? How often?

2. The cadets of the United States Military Academy at West Point, New York, are educated and supported at the expense of the United States government. The present number is 250, and the pay of each is 28 dollars per month. What is the total expense for 12 months?

Ans. 84,000 dollars.

Q. Where is the United States Military Academy? By whom are the expenses of the cadets borne? How many cadets in the Academy? What is their pay per month?

DIVISION OF SIMPLE NUMBERS.

38. *Division teaches the method of finding how many times one number is contained in another.*

The number to be divided is called the *dividend*, the number to divide by, the *divisor*, and the number of times the dividend contains the divisor, is called the *quotient*.

A quotient is *perfect* or *complete*, when the dividend contains the divisor an exact number of times; and *imperfect* when there is a number left. The number left is called the *remainder*.

In *division of simple numbers*, the numbers considered in any one example are supposed to be expressed in terms of the *same kind of unit;* that is, they are all dollars, or all pounds, or all miles.

To indicate division we use the sign ÷. Thus, 6÷3 denotes that 6 is to be divided by 3.

Q. What does *Division* teach? Which number is called the dividend? Which is the divisor? Which the quotient? When is a quotient per-

fect or complete? When imperfect? What is the number *left* called?
In division of *simple* numbers, what kind of numbers are expressed?
How do you explain this? What sign is used to indicate division?
What does 6÷3 denote?

39. If we had to divide 6 apples equally among 3 boys,
each would have 2 apples, and there would be no apples left.
Here 6 apples is the number to be divided, or *dividend*, 3
the *divisor*, and 2 the *quotient;* and as there are no apples
left, the quotient is *perfect.*

40. To divide 7 apples equally among 3 boys, each still
has 2 apples, but 1 apple is left; the quotient therefore is
imperfect, and 1 apple is the *remainder.*

To explain the operation in these
two examples, suppose that, in the
first case, we give 1 apple to each
boy; as there are 3 boys, this will
take 3 apples of the 6, and there will
be 3 apples left. Give another apple
to each of the boys; this will take 3
more, and as we had only 3 left, it
will require all that we have. Each
boy then gets 2 apples, and none are left.

OPERATION.
6
3
—
3
3
—
0 remainder.

Again, if we have 7 apples, give
as before 1 apple to each boy; this
takes 3 apples and leaves 4. Giving
another apple to each boy, 3 more
are taken from the 4 remaining, and
1 apple is left.

OPERATION.
7
3
—
4
3
—
1 remainder.

Q. In dividing 6 apples equally among 3 boys, how many does each
get? Which number is the dividend? Which the divisor? Which
the quotient? Is there any remainder? Is the quotient perfect or imperfect? Why perfect? In dividing 7 apples equally among 3 boys,
is there any remainder? What is it? What is the quotient? Is it
perfect or imperfect? Why imperfect? Explain the operation in these
two cases. How would you divide 20 apples among 5 boys? How
many would each have? Would the quotient be perfect or imperfect?
Why perfect? Is there a remainder? If we divide 21 needles among
4 girls, how many will each get? Is there a remainder? What is it?
Is the quotient perfect or imperfect? Why imperfect? Explain the
operation in the first case. Explain it in the second case.

41. In like manner we may always find how many times one number contains another, by continually subtracting the less number from the greater. To find how many times 12 is contained in 60, we find, by performing the successive subtractions, that it is contained 5 times, and there is no remainder. The quotient is therefore 5, and it is perfect.

OPERATION.

```
60
12  First Subtraction.
──
48
12  Second Subtraction
──
36
12  Third Subtraction.
──
24
12  Fourth Subtraction.
──
12
12  Fifth Subtraction.
──
 0  Remainder.
```

The above operation is tedious, when the dividend is large with respect to the divisor. We may arrive at the same result in less time by means of *Division*.

Division is therefore regarded as a short method of working subtraction.

Q. How may you find how many times one number contains another? Explain the operation of finding how many times 60 contains 12. How many subtractions are made? What then is the quotient? Is it perfect? Why? Might this method be applied to any other numbers? Why is it best to adopt another method? What other method is used? What may Division be regarded?

42. We might form, in this way, by performing the successive subtractions indicated, a *Table*, which would express the different *quotients* arising from the division of the numbers considered by the numbers 1, 2, 4, 5, &c., as we have already done for the *products* of numbers in the Multiplication Table. It would only be necessary to see how many times the less number could be subtracted from the greater, and to place the results in corresponding columns of the table. But it will be observed, if one of these quotients be known, the divisor multiplied by this quotient must produce the dividend.

For, if 3 be contained in 6, 2 times, 2 times 3 must produce 6; and if 12 be contained in 60, 5 times, 5 times 12 must produce 60.

The Multiplication Table, then, already constructed, enables us at once to find the quotients in the division of simple numbers.

Thus, 2 in 4 goes 2 times, since 2 times 2 are 4.

　　　2 in 6 goes 3 times, since 3 times 2 are 6.

　　　2 in 8 goes 4 times, since 4 times 2 are 8.

　　　3 in 6 goes 2 times, since 2 times 3 are 6.

　.　3 in 9 goes 3 times, since 3 times 3 are 9.

　　　3 in 12 goes 4 times, since 4 times 3 are 12.

　　　4 in 20 goes 5 times, since 5 times 4 are 20.

　　　4 in 28 goes 7 times, since 7 times 4 are 28.

　　　5 in 20 goes 4 times, since 4 times 5 are 20.

　　　6 in 60 goes 10 times, since 10 times 6 are 60.

　　　7 in 56 goes 8 times, since 8 times 7 are 56.

　　　8 in 72 goes 9 times, since 9 times 8 are 72.

　　　9 into 108 goes 12 times, since 12 times 9 are 108.

So that any one familiar with the Multiplication Table will promptly give the quotients arising from the division of si. milar numbers.

Q. Might a *Division Table* be constructed as we have done in the Multiplication Table? How? What would the result express? Is it necessary to make this table? Why not? How many times will 2 go in 4? Why 2 times? 2 in 6? 2 in 8? 3 in 6? 3 in 9? 3 in 12? 3 in 15? 3 in 21? 4 in 8? 4 in 12? 4 in 20? 5 in 20? 5 in 25? 6 in 12? 6 in 18? 6 in 24? 6 in 36? 7 in 49? 7 in 56? 8 in 56? 8 in 72? 9 in 72? 9 in 108? 10 in 90? 10 in 100? 11 in 88? 11 in 132? 12 in 144?

43. We distinguish two kinds of Division:

1. *Short Division,* when the divisor does not exceed 12.

2. *Long Division,* when the divisor exceeds 12.

・ *Q.* How many kinds of Division are there? What are they? What is Short Division? What is Long Division?

SHORT DIVISION—DIVISOR NOT GREATER THAN 12.

44. Divide 684 by 2.

We place the divisor 2 on the left of the dividend, and draw a curve line to separate them, and then draw a straight line under the dividend. We

OPERATION.

2) 684 dividend.

　　342 quotient.

commence on the left, and say, 2 in 6 goes 3 times: since the 6 we are dividing is *hundreds,* the quotient 3 is also *hundreds,* and must be placed under 6 in the hundreds' place; 2 in 8 tens goes 4 tens times, we place the 4 under tens; 2

in 4 units goes 2 *units* times, the 2 is placed under the units. 342 is the quotient sought.

Q. In dividing 6 hundreds by 2, is the quotient hundreds, tens, or units? In dividing 8 tens by 2, what is the quotient? In dividing 4 units by 2?

45. Divide 974 by 3.

OPERATION.

$$3)974$$
$$324—2=324\tfrac{2}{3}$$

As in the last example, place the divisor on the left of the dividend, and say, 3 in 9 goes 3 times, set down 3; 3 in 7 goes 2 times and 1 over, set down 2; as the 1 which is left is in the tens' place, it is equal to 10 units, and as these 10 units have not been divided, we add them to the 4 units, which make 14 units; 3 in 14 goes 4 times and 2 over. As there are no other numbers to divide, we set the remainder 2, a little to the right of the quotient, separating it from the quotient by a small line; or the quotient might have been expressed thus—324⅔, the remainder 2 being placed over the divisor, and a line drawn between them, to indicate the division which cannot be performed.

46. Divide 1105 by 9.

OPERATION.

$$9)1105$$
$$122—7=122\tfrac{7}{9}$$

Here we cannot divide the first number 1 by 9, but as this figure occupies the place of thousands, the 1 thousand is the same as 10 hundreds, which being added to 1 hundred, make 11 hundreds; then 9 in 11 hundreds goes 1 hundred times, and 2 hundreds over, set down 1 and carry 2 hundreds, or 20 tens to 0 tens, which makes 20 tens; 9 in 20 tens goes 2 tens' times and 2 tens over; carrying the 2 tens or 20 units to the 5 units, we have 25 units; 9 in 25 units goes 2 units times and 7 over. The quotient sought is 122⅞.

47. We see by the above examples that if units be divided by a simple number, the quotient will be units; if tens be divided by the same number, the quotient will be tens; if hundreds be divided, the quotient will be hundreds, &c.

Q. How are the numbers arranged in short division? Where do you commence to divide? In dividing 6 hundreds by 2, what is the quotient? If you divide 8 tens by 2? 4 units by 2? In dividing 7 tens by 3, what do you do with the 1 ten that is over? 1 ten is equal to how many units? If there is a remainder after division, how is it written?

From these illustrations we deduce the following

RULE.

48. I. *Set down the divisor on the left of the dividend, draw a curve line between them, and a straight line under the dividend.*

II. *Commencing on the left, find how many times the divisor is contained in the first or first two figures of the dividend. Place the number so found under the dividend as the first figure of the quotient, observing that this figure must occupy the units' place, if units have been divided, tens' place if tens, &c.*

III. *If there is no remainder, divide the next figure in the dividend by the divisor, set down the figure found in the quotient, and proceed in this way until all the figures in the dividend are divided.*

IV. *But should there be a remainder in dividing any figure in the dividend, multiply this remainder by 10 and add the product to the next figure in the dividend. This sum divided by the divisor will give the next figure in the quotient.*

PROOF FOR DIVISION.

49. *Multiply the quotient by the divisor and add to this product the remainder, if there be any. The result will be equal to the dividend, if the work is right.*

Divide 846321 by 6, and 10432 by 2.

```
6)846321                    2)10432
  --------                    -------
  141053—3                    5216
       6                         2
  --------                    -------
  846318                      10432
     3 remainder.
  --------
   846321
```

Multiplying the quotient 141053 by the divisor 6, the product is 846318, to which adding the remainder 3, the sum is the same as the dividend. In the second example, there being no remainder, the quotient 5216 multiplied by the divisor 2 produces the dividend.

Q. Repeat the rule for short division. What do you do when you have a remainder? What is the proof for division?

EXAMPLES.

1. Divide 100457 by 2.   *Ans.* 50228½.

2. Divide 114732 by 3.   *Ans.* 38244.

3. Divide 94371407 by 4.   *Ans.* 23592851¾.

4. Divide 20043041 by 5.   *Ans.* 4008608⅕.

5. Divide 876432189 by 6.   *Ans.* 146072031½.

6. Divide 184652417 by 7.   *Ans.* 26378916⅝.

7. Divide 98434899 by 8.   *Ans.* 12304362⅜.

8. Divide 488766789 by 9.   *Ans.* 54307421.

9. Divide 104763040 by 10.   *Ans.* 10476304.

11. Divide 121763240 by 11.   *Ans.* 11069385$\frac{5}{11}$.

12. Divide 897653218 by 12.   *Ans.* 74804434$\frac{10}{12}$.

APPLICATIONS.

1. The distance of the sun from the earth is computed at 95,000,000 of miles, and light is 8 minutes in coming from the sun to the earth; how many miles does light travel in 1 minute?   *Ans.* 11,875,000 miles.

Q. How far is the sun from the earth? How many minutes is light in coming from the sun to the earth?

2. The principal rivers in the United States are the Mississippi, which is 4100 miles long; the Missouri, which is 2900 miles long; the Arkansas, 2000; Ohio, 1300; Tennessee, 900; Alabama, 600; Potomac, 500; James, 450; Hudson, 350. How long would a steam-boat take to travel each of these rivers, at the rate of 7 miles per hour?

     Ans.

Q. What are the principal rivers in the United States? Which is the longest? How many miles in length? Which is the next in length? How long? How long is the James River? The Hudson? The Ohio? &c. &c.

C

3. What is the monthly pay of the Governors of the diferent States, their annual salaries being as follows:

States.	Governor.	Salary. Dollars.	Monthly Pe Dollar
1. Maine, - -	H. J. Anderson,	$1500 \div 12 = Ans.$ 12
2. N. Hampshire,	J. Steele, - -	$1200 \div 12 = Ans.$ 10
3. Vermont, -	John Mattocks,	$750 \div 12 = Ans.$
4. Massachusetts,	G. N. Briggs, -	$2500 \div 12 = Ans.$
5. Rhode Island,	James Fenner,	$400 \div 12 = Ans.$
6. Connecticut,	R. Baldwin, -	$1000 \div 12 = Ans.$
7. New York, -	Wm. C. Bouck,	$4000 \div 12 = Ans.$
8. New Jersey,	Daniel Haines,	$2000 \div 12 = Ans.$
9. Pennsylvania,	David R. Porter,	$4000 \div 12 = Ans.$
10. Delaware, -	Wm. B. Cooper,	$1333 \div 12 = Ans.$
11. Maryland, -	Francis Thomas,	$4200 \div 12 = Ans.$
12. Virginia, - -	James M'Dowell,	$3333 \div 12 = Ans.$
13. N. Carolina,	J. M. Morehead,	$2000 \div 12 = Ans.$
14. S. Carolina,	J. H. Hammond,	$3500 \div 12 = Ans.$
15. Georgia, - -	G. W. Crawford,	$3000 \div 12 = Ans.$
16. Alabama, -	Benj. Fitzpatrick,	$2500 \div 12 = Ans.$
17. Mississippi, -	Albert G. Brown,	$3000 \div 12 = Ans.$
18. Louisiana, -	Alex. Mouton,	$7500 \div 12 = Ans.$
19. Arkansas, -	Archibald Yell,	$1800 \div 12 = Ans.$
20. Tennessee, -	James C. Jones,	$2000 \div 12 = Ans.$
21. Kentucky, -	Robt. P. Letcher,	$2500 \div 12 = Ans.$
22. Ohio, - - -	T. W. Bartley,	$1500 \div 12 = Ans.$
23. Michigan, -	John S. Barry,	$1500 \div 12 = Ans.$
24. Indiana, - -	James Whitcomb,	$1500 \div 12 = Ans.$
25. Illinois, - -	Thomas Ford,	$2000 \div 12 = Ans.$
26. Missouri, - -	Thos. Reynolds,	$2000 \div 12 = Ans.$

TERRITORIES.

Florida, - - -	John Branch, -	$2500 \div 12 = Ans.$
Wisconsin, - -	N. P. Tallmadge,	$2500 \div 12 = Ans.$
Iowa, - - - -	J. Chambers, -	$2500 \div 12 = Ans.$

Q. How many States are there in the United States? What ai they? How many Territories? What are they? Who is the G(vernor of your State? What is his salary? Which State gives tl *largest* salary to its Governor? Which the *smallest*?

LONG DIVISION—DIVISOR GREATER THAN 12.

49. Divide 6314 by 13.

We commence by placing the divisor as in Short Division, and draw a curve line on the right of the dividend, to separate it from the quotient. We cannot divide 6 by 13, but 13 in 63 goes 4 times and something over; place 4 in the quotient; it is hundreds, since 63 hundreds divided by 13 give hundreds. Multiply 13 by 4, the product is 52, which being taken from 63 hundreds leaves 11 hundreds; 4 is then the greatest number of times that 63 contains 13, since the remainder 11 is less than 13. The remainder being 11 hundreds, it is equal to 110 tens, and bringing down 1 ten in the dividend, the next number to divide is 111 tens; 13 in 111 goes 8 times and something over. As the number divided is tens, the quotient figure 8 is set in the tens' place. Multiply 13 by 8, the product is 104, which being subtracted from 111 leaves 7 for a remainder; the quotient 8 is not too large, since the remainder is less than the divisor. Bring down the next figure 4 of the dividend; 74 units contain 13, 5 times and something over; placing 5 in the units' place in the quotient, and multiplying the divisor by it, the product 65 subtracted from 74 leaves 9 for the last remainder. The quotient sought is $485\frac{9}{13}$.

OPERATION.

divisor. dividend. quotient.

$$13 \,)\, 6314 \,(\, 485\tfrac{9}{13}.$$
$$52$$
$$\overline{111}$$
$$104$$
$$\overline{\cdot\cdot 74}$$
$$65$$
$$\overline{\cdot 9 \text{ remainder.}}$$

50. Divide 87525 by 25.

We say 25 in 87 goes 3 times, place 3 in the quotient. Multiplying 25 by 3 and subtracting the product, we have 12 for a remainder. Bringing down 5, we find the divisor is contained in 125, 5 times and no remainder; set 5 in the quotient. Bring down 2, here 2 contains 25, 0 time, set 0 in the quotient and bring down 5; 25 contains 25, 1 time and no

OPERATION.

$$25 \,)\, 87525 \,(\, 3501$$
$$75$$
$$\overline{125}$$
$$125$$
$$\overline{\cdot\cdot\cdot 25}$$
$$25$$
$$\overline{00 \text{ remainder.}}$$

remainder; setting 1 in the quotient, the dividend contains the divisor 3501 times.

51. *Note.*—We may more readily ascertain the number of times the dividend contains the divisor, by finding how many times the first figure in the divisor is contained in the first or first two figures of the dividend. Should this number be too large, diminish it by 1 until the divisor multiplied by it will give a product less than the partial dividend. Thus,

Divide 1755 by 39.

OPERATION.

The first two figures 17 contain 3, 5 times, but 5 times 39 are 195, which being greater than 175, the quotient figure 5 is too large; make it 4: 4 times 39 are 156, and the remainder 19 being less than 39, 4 is the first figure of the quotient. Again, 3 in 19 goes 6 times, but 6 times 39 being 234, we try 5 for the second figure in the quotient, and find that 5 times 39 are exactly equal to 195. The quotient is therefore 45.

$$39) 1755 (45$$
$$156$$
$$\overline{\cdot 195}$$
$$195$$
$$\cdots$$

Q. In dividing the first or first two figures of the dividend by the first figure of the divisor, how do you know when the quotient figure is too large? How do you get the true quotient figure when it is too large? Can the remainder be larger than the divisor if the quotient figure be right? Can it be equal to it?

From these illustrations we deduce the following

RULE.

52. I. *Set the divisor on the left of the dividend, and draw a curve line between them. Draw another curve line on the right of the dividend, to separate it from the quotient.*

II. *Take on the left hand of the dividend as many figures as will contain the divisor, as a partial dividend. Find how often the first figure in the divisor is contained in the first or first two figures of this partial dividend. Multiply the whole divisor by the number so found, and if the product is less than the partial dividend, the quotient figure is not too large. Set this figure on the right of the dividend as the first figure in the quotient. But if this product is greater than the partial dividend, diminish the quotient figure by 1, until the product of the divisor by it is less than the partial dividend.*

III. *Subtract the product of the divisor by the quotient figure found from the partial dividend, and to the remainder annex the next figure in the dividend for a new dividend.*

IV. *Divide the first or first two figures of this remainder by the first figure in the divisor, diminishing the quotient figure by 1, if necessary, as before, for the next figure in the quotient. Multiply the whole divisor by it, and subtract this product from the new dividend. Bring down the next figure of the dividend, and continue this operation until all the figures of the dividend are brought down.*

PROOF FOR LONG DIVISION.

53. *Multiply the quotient by the divisor, and to this product add the remainder, if there be any; the sum will be equal to the dividend, if the work is correct.*

Divide 87432 by 253.

OPERATION.	PROOF.
253) 87432 (345	345 quotient.
759	253 divisor.
1153	1035
1012	1725
1412	690
1265	87285
147	147 remainder.
	87432 dividend.

The sum 87432 being the same as the dividend, the work is correct.

Q. Repeat the rule for Long Division. How are the numbers set down? Where is the quotient placed? How many figures do you take in the dividend as a partial dividend? How then do you proceed? When you divide the first or first two figures of this partial dividend by the first figure of the divisor, will the quotient always be the true quotient? If it is too large what do you do? Where do you place this quotient figure? What do you do with it next? What is the next step? If the remainder you get is larger than the divisor, is the quotient figure too large or too small? What do you annex to this remainder? How long do you continue this operation? What is the proof for Long Division? Does it differ from the proof for Short Division?

EXAMPLES.

1. Divide 136704 by 256.   *Ans.* 534.
2. Divide 666490 by 365.   *Ans.* 1826.
3. Divide 105062176 by 1208.   *Ans.* 86972.
4. Divide 151807361 by 12321.   *Ans.* $12321\frac{320}{12321}$.
5. Divide 808206 by 2048.   *Ans.* $394\frac{1294}{2048}$.
6. Divide 2857431 by 751.   *Ans.* $3804\frac{627}{751}$.
7. Divide 27097441434 by 27639.   *Ans.* 980406.

APPLICATIONS.

1. In 1840, there were 3,242 Academies and Grammar-Schools in the United States, containing 164,159 students. What was the average number of students in each ?

Ans. $50\frac{2042}{3242}$ students.

Q. How many Academies and Grammar-Schools in the United States ? How many students in all ? What is the average number in each ?

2. What is the daily pay of the President of the United States, his annual salary being 25000 dollars, and the year being supposed to contain 365 days.   *Ans.* $68\frac{180}{365}$ dollars.

Q. What is the salary of the President of the United States ?

3. What was the average annual number of emigrants to the United States from Great Britain and Ireland, from 1825 to 1837 inclusive, the number each year being as follows :

Number of Emigrants.

1825	5,551
1826	7,063
1827	14,526
1828	12,917
1829	15,678
1830	24,887
1831	23,418
1832	32,872
1833	29,109
1834	33,074
1835	26,720
1836	37,774
1837	36,770

$\div$ 13 = *Ans.* $23,096\frac{11}{13}$.

4. What is the average number of white persons over the age of 20 years in each county of the following States who cannot read and write, the total number of white persons who cannot read and write, and the number of counties in each state being as follows, in 1840.

States.	No of persons over 20 yrs. who cannot read or write.	No. of counties.	No. of persons in each county who cannot read or write.
1. Maine	3,241 ÷	13 =	Ans. 249 $\frac{4}{13}$.
2. New Hampshire .	942 ÷	9 =	Ans. 104 $\frac{2}{3}$.
3. Massachusetts ...	4,448 ÷	14 =	Ans. 317 $\frac{5}{7}$.
4. Rhode Island....	1,614 ÷	5 =	Ans. 322 $\frac{4}{5}$.
5. Connecticut	526 ÷	8 =	Ans. 68 $\frac{3}{4}$.
6. Vermont	2,270 ÷	10 =	Ans. 227.
7. New York......	44,452 ÷	58 =	Ans.
8. New Jersey.....	6,385 ÷	18 =	Ans.
9. Pennsylvania ...	33,940 ÷	54 =	Ans.
10. Delaware.......	4,832 ÷	3 =	Ans.
11. Maryland	11,817 ÷	20 =	Ans.
12. Virginia	58,787 ÷	119 =	Ans.
13. North Carolina..	56,602 ÷	68 =	Ans.
14. South Carolina ..	20,615 ÷	29 =	Ans.
15. Georgia	30,717 ÷	93 =	Ans.
16. Alabama	22,592 ÷	49 =	Ans.
17. Mississippi......	8,360 ÷	56 =	Ans.
18. Louisiana	4,861 ÷	38 =	Ans.
19. Tennessee......	58,531 ÷	72 =	Ans.
20. Kentucky	40,018 ÷	90 =	Ans.
21. Ohio	35,394 ÷	79 =	Ans.
22. Indiana	38,100 ÷	87 =	Ans.
23. Illinois.........	27,502 ÷	87 =	Ans.
24. Missouri	19,457 ÷	62 =	Ans.
25. Arkansas	6,567 ÷	39 =	Ans.
26. Michigan.......	2,173 ÷	32 =	Ans.

Q. Which State has the greatest number of white persons over 20 years of age who cannot read or write? How many such has Virginia? Which State has the smallest number? How many in Connecticut? How many in your State over 20 years of age who can neither read nor write? How many counties has your State? How many such persons in each county of your State?

5. By the census of 1840, there are 173 Universities and Colleges in the United States, containing in all 16,233 students. What is the average number of students in each?

Ans. 93$\frac{144}{173}$ students.

Q. How many Colleges or Universities are there in the United States? What is the total number of students? How many does each average?

54. REMARK I. *When the divisor is* 10, 100, 1000, 10000, &c., the division may be immediately effected by cutting off from the right hand of the dividend as many figures as there are ciphers in the divisor, the figures cut off when they are significant being placed over the divisor and connected with the quotient. Thus,

Divide 8846 by 100.

There being two ciphers in the divisor, two figures, 4 and 6, are cut off, the quotient is therefore 88$\frac{46}{100}$. The reason of this is evident, since cutting off 1, 2, 3, &c. figures from the dividend, diminishes it 10, 100, 1000, &c. times.

OPERATION.

1,00) 88,46

88$\frac{46}{100}$ quotient.

Q. When the divisor is 10, 100, 1000, &c., how may the division be at once effected? How many figures are cut off when the divisor is 10? 100? 1000? 10000? What is done with the significant figures which are cut off? What is the quotient of 800 divided by 10? Divided by 100? What is the reason of this rule? How many tens in 800? How many hundreds?

55. REMARK II. *When there are ciphers on the right of the divisor,* we may simplify the division by cutting off as many figures in the dividend as there are ciphers in the divisor, then dividing the remaining figures of the dividend by the significant figures of the divisor, annexing to the remainder, if there be any, the figures cut off in the dividend. Thus,

Divide 8246 by 500.

In this example we cut off two figures in the dividend and the two ciphers in the divisor, and dividing 82 by 5, we find a quotient 16 and 2 over. As this 2 remains from the division of the 2 *hundreds* the dividend, it is also hundreds, and is therefore placed

OPERATION.

5,00) 82,46

16$\frac{246}{500}$

on the left of the figures cut off, which contain only tens and units. The true remainder is therefore 246.

For, cutting off two figures from the dividend is the same thing as dividing the number by 100 ; and if we divide the number after the two figures are cut off, by 5, we shall find the 500th part of the given number.

Q. When there are ciphers on the right of the divisor, how may the division be simplified ? What is done with the remainder arising from the division by the significant figures of the divisor ? Why is it placed on the left of the figures cut off in the dividend ? Cutting off two figures from the right of a number is equivalent to dividing by what ? What part of the given number will the number then express ? If this number be now divided by 5, what part of the given number will the quotient be ? If divided by 2 instead of 5 ?

56. REMARK III. When the divisor can be separated into factors, each of which is less than 12, the division may be abridged by dividing by each of these factors separately, as in Short Division. When there are remainders from the partial divisions, the true or whole remainder is found by multiplying the last partial remainder by the preceding partial divisor, and to this product adding the preceding remainder. Multiply this sum by the next preceding divisor. and add the next preceding remainder, and so on until all the remainders are considered. The result will be the true remainder. Thus,

Divide 31046835 by 56, the factors of which are 7 and 8.

OPERATION.

$$7 \,)\, 31046835$$
$$8 \,)\, 4435262 \quad - \; 1, \text{ First remainder.}$$
$$554407 \quad - \; 6, \text{ Second remainder.}$$

Ans. 554407$\frac{43}{56}$

$$
\begin{array}{l}
6 \text{ last remainder.} \\
7 \text{ preceding divisor.} \\
\hline
42 \\
1 \text{ preceding remainder.} \\
\hline
43 \text{ true remainder.}
\end{array}
$$

Or, $6 \times 7 + 1 = 43.$

c 2

Dividing first by the factor 7, the quotient is 4485262, with a remainder 1. Dividing this quotient by the other factor 8, we get a quotient 554407, and a remainder 6. To find the true remainder, we multiply the last remainder 6 by the preceding divisor 7, and to the product 42 add the preceding remainder, which is 1 ; their sum 43 is the true remainder, and the true quotient is $554407\frac{43}{56}$.

Q. When the divisor can be separated into factors less than 12, how may the division be abridged ? How may the true or whole remainder be found ? If the divisor were 56, by what would you divide ? If 81 ? If 100 ? If 121 ? If 144 ?

EXAMPLES UPON THE REMARKS.

1. Divide 1200 by 10.   *Ans.* 120.

2. Divide 847 by 100.   *Ans.* $8\frac{47}{100}$.

3. Divide 321804 by 1000.   *Ans.* $321\frac{804}{1000}$.

4. Divide 4721 by 40.   *Ans.* $118\frac{1}{40}$.

5. Divide 82053 by 700.   *Ans.* $117\frac{153}{700}$.

6. Divide 1005405 by 3500.   *Ans.* $287\frac{905}{3500}$.

7. Divide 1870369 by 809000.   *Ans.* $2\frac{252369}{809000}$.

8. Divide 7048321 by 72, the factors of which are 9 and 8.   *Ans.* $97893\frac{25}{72}$.

9. Divide 100325 by 96, the factors of which are 12 and 8.   *Ans.* $1045\frac{5}{96}$.

10. Divide 5130652 by 132, the factors of which are 12 and 11.   *Ans.* $38868\frac{76}{132}$.

11. Divide 984706 by 144, the factors of which are 12 and 12.   *Ans.* $6838\frac{34}{144}$.

12. Divide 327605 by 210, the factors of which are 5, 6, and 7.   *Ans.* $1560\frac{5}{210}$.

13. Find the three factors of 504, and divide 89760301 by them.   *Ans.* $178095\frac{421}{504}$.

14. Find the three factors of 160, and divide 298305602 by them.   *Ans.* $1864410\frac{2}{160}$.

APPLICATIONS UPON THE PRECEDING RULES.

1. What is the annual expense of the United States Su-*eme Court* which consists of a Chief Justice, who receives

5000 dollars, and eight Associate Justices, who receive 4500 dollars each? *Ans.* 41000 dollars.

Q. What officers compose the United States Supreme Court? What is the salary of the Chief Justice? What of each of the Associate Justices?

2. What is the annual expense of the Cabinet of the President of the United States, which consists of a *Secretary of State*, a *Secretary of the Treasury*, a *Secretary of War*, a *Secretary of the Navy*, and a *Postmaster-General*, each of whom receives 6000 dollars per year, and an *Attorney-General*, who receives 4000 dollars?

Ans. 34000 dollars.

Q. What officers compose the Cabinet of the President of the United States? Do they all receive the same salary? Who receives the least?

3. The government of the United States is represented at the courts of *Great Britain, France, Russia, Prussia, Austria, Spain, Brazil* and *Mexico*, by an agent called a *Minister Plenipotentiary*, each of whom receives an annual salary of 9000 dollars. Each minister is allowed a Secretary, with a salary of 2000 dollars. What is the total annual expense of these officers? *Ans.* 88000 dollars.

Q. What is the agent called that represents the United States at the Court of Great Britain? How many Courts are thus represented? What are they? What salary is allowed to a minister? Has he a secretary? At what salary?

4. How much does the present debt of Pennsylvania exceed the estimated value of her public improvements, her debt and the estimated value of her property being as follows, in 1843 :

How contracted.	*Debt. Dollars.*	*Public Property.*	*Estimated val. Dollars.*
Canals and Railways.....	30,533,629	Public Improvements...	30,533,629
To pay interest on debt...	4,410,135	Bank Stock............	2,106,700
For the use of Treasury..	1,571,689	Turnpike, &c. Stock....	2,836,262
Turnpikes and Roads.....	930,000	Canal, &c. Stock	842,776
Union Canal	200,000	Railroad Stock.........	365,276
Eastern Penitentiary.....	120,000	Money due	1,000,000
Franklin Railroad........	100,000		
Pennsylvania and Ohio Canal.................	50,000		
Insane Asylum..........	22,335		

Ans. 351,143 dollars.

Q. How was the debt of Pennsylvania contracted? What is the nature of its public property?

5. What is the daily expense of the governments of the following States, the total amount paid per year for each object of expenditure being as represented in the accompanying table?

Amount paid for

States.	State Officers.	Judiciary.	Legislature.	Prisons and Insane.	Militia.	Printing.	
	D.	D.	D.	D.	D.	D.	D.
Maine	6,576+	11,200+	40,687+	26,409+	8,325+	2000+365=*Ans.* 260 297/365
N. Hamp..	3,000+	13,003+	20,183+	8,000	.....	.....+365=*Ans.*
Vermont ..	2,275+	7,325+	21,003+	5,096	.....	.....+365=*Ans.*
Massachu't	16,500+	50,536+	114,071+	20,506+	21,033+	11.944+365=*Ans.*
R. Island..	1,800+	1,750+	5,346+	6,520+	.....+	634+365=*Ans.*
Connect ..	4,734+	5,650+	11,840+	......	.....	.....+365=*Ans.*
New York.	42,321+	35,128+	124,026+	155,520+	18,171+	28,241+365=*Ans.*
New Jersey	7,840+	5,880+	18,869+	16,590+	1,065+	2,973+365=*Ans.*
Pennsylv'a	39,773+	107,600+	135,988+	23,195+	31,738+	60,448+365=*Ans.*
Maryland.	14,682+	40,522+	60,632+	41,777+	2,801	.....+365=*Ans.*
Delaware .	2,733+	4,400+	14,580+	......	.....	.....+365=*Ans.*
Virginia ..	19,433+	44,550+	95,056+	64,015+	59,237	.....+365=*Ans.*
N.Carolina	6,300+	24,834+	49,620+	......	.....	0,660+365=*Ans.*
S. Carolina	15,100+	33,000+	43,520+	.....	13,714+	5,129+365=*Ans.*
Georgia ...	12,000+	21,000+	91,500+	......	.....	.....+365=*Ans.*
Alabama..	7,850+	33,775+	46,000+	......	.....	8,000+365=*Ans.*
Mississippi	10,259+	32,000+	34,552+	51,018+	703	.....+365=*Ans.*
Louisiana.	26,000+	59,000+	31,220+	......	.....	.....+365=*Ans.*
Tennessee.	8,779+	34,482+	34,171+	14,319+	.....+	4,950+365=*Ans.*
Kentucky..	10,150+	34,900+	36,746+	33,479	....	.....+365=*Ans.*
Ohio......	16,347+	24,202+	48,830+	49,239+	2,832+	20,340+365=*Ans.*
Indiana....	3,850+	20,078+	45,052+	19,651+	.....+	12,457+365=*Ans.*
Illinois....	4,550+	13,000+	63,641+	......	.....	23,845+365=*Ans.*
Missouri...	8,700+	17,050+	25,200+	.....	.....	.....+365=*Ans.*
Michigan..	5,100+	7,100+	24,962+	17,799+	....	10,406+365=*Ans.*
Arkansas.	5,200+	13,800+	17,573+	19,596+	.....	.....+365=*Ans.*

Q. What are the objects of State expenditures? What is the amount paid to each object in your state?

6. Find the total amount of debt, and the average of debt to each inhabitant of the following countries, the total number of inhabitants, and the debt of each country, being as follows:

Country.	Debt. Dollars:	No. of Inhabitants.	Average each.
England....	5,556,000,000÷	25,300,000=	*Ans.*
France	1,800,000,000÷	33,600,000=	*Ans.*
Holland	800,000,000÷	2,820,000=	*Ans.*
Denmark ...	93,000,000÷	2,097,400=	*Ans.*
Greece	44,000,000÷	810,000=	*Ans.*
Portugal....	142,000,000÷	3,400,000=	*Ans.*
Spain	467,000,000÷	11,963,000=	*Ans.*
Austria	380,000,000÷	34,100,000=	*Ans.*
Russia ..	150,000,000÷	51,100,000=	*Ans.*

Q. Do other countries appear to be in debt as well as the United States? What proportion does the proportionate debt of the States bear to that of England? France?

7. By the census of 1840, the number of inhabitants in the United States was 17,063,353. Find the total amount of the debts of the States, and the amount to each individual, the indebtedness of each State being as follows:

States.	Debt.
Maine	1,678,367
Massachusetts......	5,149,137
New York	20,165,254
New Jersey	83,283
Pennsylvania	34,723,261
Maryland	15,109,026
Virginia	6,857,161
South Carolina.....	3,764,734
Georgia	500,000
Alabama..........	10,859,556
Mississippi	12,400,000
Louisiana	20,585,000
Tennessee	1,789,166
Kentucky	4,665,000
Ohio.............	14,809,476
Indiana...........	13,667,433
Illinois	13,465,682
Missouri	2,929,557
Michigan	6,011,000
Arkansas	3,755,362
Florida (Ter.)	3,900,000
District of Columbia.	1,500,000

$\div 17,063,353 =$

Ans. $11\frac{10670572}{17063353}$ dollars to each individual.

OF FRACTIONS.

57. *If the unit 1 be divided into any number of equal parts, one or more of these parts is called a fraction.*

If the unit 1 be divided into two equal parts, one of these parts is called *one half;* two of them *two halves;* three of them *three halves;* &c.   *One half, two halves, three halves, four halves,* &c., are *fractions.*

As the unit 1 has been divided into *two* equal parts called halves, there are *two halves in* 1, *four* halves in 2, *six* halves in 3, eight halves in 4, &c. : the number of halves in every case being found by multiplying the number by 2.

Q. If the unit 1 be divided into any number of equal parts, what is one or more of these parts called ?   If the unit be divided into two equal parts, what is one of these parts called?   What are two of them called?   Three ?   Four ?   What is one half?   Two halves ?   Three halves ?   How many halves in 1?   Why ?   How many in 2 ?   In 3 ?   In 4 ?   In 5 ?   In 6 ?   How do you find the number of halves in a number ?   How many halves in 100 ?   In 150 ?   In 225 ?

58. If the unit 1 be divided into three equal parts, these parts are called *thirds,* and one of these parts is *one-third;* two of them *two-thirds;* three, *three-thirds;* four, *four-thirds,* &c.   One-third, two-thirds, &c., are also *fractions.*

The unit 1 being divided into *three equal* parts, there are three thirds in 1, six thirds in 2, nine thirds in 3, twelve thirds in 4, &c.   The number of thirds in any number being found by multiplying the number by 3.

Q. If the unit be divided into three equal parts, what are these parts called ?   What is one of them called ?   Two ?   Three ?   What is one-third ?   Why a fraction ? .   What is two-thirds ?   How many thirds in 1?   Why?   In 2 ?   In 3 ?   In 4 ?   How do you find the number of thirds in any number ?   How many thirds in 50 ?   In 100 ?   In 150 ?

59. When the unit 1 is divided into *four, five, six,* &c. equal parts, these parts are called *fourths, fifths, sixths, sevenths,* &c., and one fourth, two fourths, one fifth, two fifths, one sixth, two sixths, one seventh, two sevenths, three sevenths, &c., are fractions.

There are then four fourths in 1, five fifths in 1, six sixths in 1, &c., since the unit is divided into 4, 5, 6, 7, &c. *equal* parts.

There being four fourths in 1, there must be eight *rths in* 2, twelve fourths in 3, &c.   The number of fourths in the number being equal to the number multiplied by 4. Thus there are 20 fourths in 5, since 4 × 5 gives 20. In like manner there being five fifths in 1, there will be ten fifths in 2; and there being six sixths in 1, there will be 12 sixths in 2, &c. Hence we conclude, that to find the number of fifths, sixths, sevenths, &c., in any number, we multiply by 5 to find the fifths, by 6 to find the sixths, by 7 to find the sevenths, &c.

To find the number of fifths in 20, we multiply 20 by 5, and we get 100; there are then 100 fifths in 20. There are 200 tenths in 20, since 20 × 10 gives 200.

Q. When 1 is divided in 4 equal parts, what are these parts called? What when divided into 5 equal parts? Into 6? How is the unit divided in one-fifth? In two sevenths? What are one fifth, two fifths, one sixth called? How many fifths in one fifth? In three fifths? How many sixths in four sixths? How many fourths in 1? How many fifths? Sixths? How many fourths in 2? In 3? In 4? How do you find the number of fourths in a number? How many fifths in 1? How many sixths? How many sevenths? How many tenths? How many fifths in 2? How many sixths? How many tenths? How many fifths, sixths, sevenths, tenths, &c. in 3? In 4? How do you find the number of fifths in a number? The number of sixths? The number of tenths? How many fifths in 20? How many tenths? How many in 50? How many fifths in ten fifths?

60. A fraction is expressed by two numbers written one above the other, with a line drawn between them; the lower number shows into how many equal parts the unit 1 is divided, and the upper number how many of these parts are expressed in the fraction.

Thus, in the fraction *one-half*, the unit being divided into *two* equal parts, the lower number is 2; and since one of these parts is taken, the upper is 1; then *one-half* is written $\frac{1}{2}$. In the fraction *three halves*, the unit is again divided into *two* equal parts, but *three* of these parts are expressed. The fraction is written $\frac{3}{2}$. In like manner we may express by figures,

Four halves, which is written, $\frac{4}{2}$
Two-thirds, " " $\frac{2}{3}$
Four-fifths, " " $\frac{4}{5}$
Five-sixths, " " $\frac{5}{6}$
Two-tenths, " " $\frac{2}{10}$
Nine-twentieths, " " $\frac{9}{20}$

The *two numbers* which compose a fraction, are called *its terms.*

The lower number is called the *denominator* of the fraction, and it *shows into how many equal parts the unit* 1 *is divided.*

The upper number is called the *numerator* of the fraction, and it *shows how many of the equal parts are expressed in the fraction.*

In the fraction $\frac{1}{2}$, 1 is the numerator and 2 the denominator.

In the fraction $\frac{4}{3}$, 4 is the numerator and 3 the denominator.

In the fraction $\frac{3}{4}$, 3 is the numerator and 4 the denominator.

Q. How is a fraction expressed? What does the lower number show? The upper? How is one-half expressed? Why is the lower number 2, and the upper number 1? How is the fraction two halves expressed? Three halves? Two thirds? What are the two numbers called which compose a fraction? Which is the numerator? What does the numerator show? Which is the denominator? What does it show? Which is the numerator and which the denominator in the fraction $\frac{2}{3}$? In $\frac{3}{3}$? Into how many equal parts is the unit divided in $\frac{2}{3}$? In $\frac{3}{3}$? How many parts are taken in $\frac{2}{3}$? How is the unit divided in the fraction $\frac{8}{13}$, and how many parts expressed? In $\frac{1}{16}$, $\frac{2}{13}$, $\frac{4}{7}$, $\frac{7}{8}$, $\frac{9}{10}$, $\frac{10}{9}$? &c.

61. A whole number may be expressed like a fraction, by writing 1 below it as its denominator. Thus 3 expressed fractionally, is $\frac{3}{1}$. Which is read 3 *ones.* Since 3 ones make 3, the value of the number is not altered by writing 1 under it as a denominator.

Q. How may a whole number be expressed fractionally? How may 3 be expressed as a fraction? How do you read $\frac{3}{1}$? Is the value of the number changed? Why not? Write 4, 5, 6, &c., in a fractional manner.

62. A fraction denotes division, and its value is equal to the quotient obtained by dividing the numerator by the denominator. Thus, $\frac{2}{2}$ is equal to 1; and since two halves make 1, the *value* of the fraction is represented by the quotient 1. In like manner, $\frac{4}{2}$ is equal to 2, $\frac{12}{3}$ to 4, $\frac{16}{4}$ to 4, $\frac{20}{5}$ to 4, &c.

From this we conclude, that if the numerator be less than the denominator, the value of the fraction is less than 1. If the numerator be equal to the denominator, the value of the fraction is 1 ; and if greater than the denominator, the value of the fraction is greater than 1.

Q. What does a fraction denote? What is its value equal to? What is the value of the fraction $\frac{3}{4}$? $\frac{4}{4}$? $\frac{10}{4}$? $\frac{8}{4}$? $\frac{16}{4}$? When the numerator is equal to the denominator, what is the value of the fraction? When greater? When less?

63. When the unit is divided into tenths, hundredths, thousandths, &c., the resulting fractions are called *Decimal Fractions.* All other fractions are called *Vulgar* or *Common Fractions.*

$\frac{1}{10}$, $\frac{2}{10}$, $\frac{1}{100}$, $\frac{3}{100}$, $\frac{1}{1000}$, $\frac{1}{1000}$, $\frac{3}{1000}$, $\frac{20}{10000}$, &c., are decimal fractions.

$\frac{2}{3}$, $\frac{1}{4}$, $\frac{1}{2}$, $\frac{3}{8}$, $\frac{5}{6}$, &c. &c., are vulgar fractions.

The denominator of a decimal fraction is not usually expressed, and to distinguish the numerator, which is alone written, from a whole number, a *point* . is placed on its left, called the *decimal point.*

$\frac{2}{10}$ is written .2, $\frac{5}{10}$ is written .5, $\frac{25}{100}$ is written .25.

When the denominator of a decimal fraction is expressed, it is always 1, with as many cyphers annexed as there are figures in the numerator.

Q. What are decimal fractions? What are other fractions called? What kind of a fraction is $\frac{1}{10}$? $\frac{3}{4}$? $\frac{1}{100}$? $\frac{5}{5}$? Are the denominators of decimal fractions usually expressed? How is the fraction written? When the denominator is expressed, what is it? What is .2? .5? .25? .05? .225? .2004? .100?

OF VULGAR FRACTIONS.

64. There are five kinds of Vulgar Fractions, viz: *Proper, Improper, Simple, Compound* and *Mixed.*

A *Proper Fraction* is one in which the numerator is less than the denominator; thus, $\frac{1}{2}$, $\frac{3}{4}$, $\frac{5}{6}$, $\frac{7}{8}$, &c., are proper fractions. A proper fraction is always less than 1.

An *Improper Fraction* is one in which the numerator is equal to or greater than the denominator; thus, $\frac{2}{2}$, $\frac{3}{2}$, $\frac{4}{3}$, $\frac{6}{5}$, are improper fractions.

A *Simple Fraction* is a single fraction, in which there is but one numerator and one denominator. A simple fraction may be either *proper or* improper. $\frac{1}{2}$, $\frac{3}{4}$, $\frac{3}{8}$, $\frac{3}{4}$, $\frac{4}{3}$, &c., are simple fractions.

A *Compound Fraction* is a fraction of a fraction. Thus the following are compound fractions :

$$\tfrac{1}{2} \text{ of } \tfrac{3}{4}; \ \tfrac{1}{3} \text{ of } \tfrac{3}{5}; \ \tfrac{1}{4} \text{ of } \tfrac{5}{6} \text{ of } \tfrac{2}{7}.$$

A *Mixed Number* is composed of a whole number and a fraction. The following are mixed numbers :

$$2\tfrac{1}{4}, \ 3\tfrac{2}{3}, \ 4\tfrac{5}{6}, \ 7\tfrac{1}{5}, \ 9\tfrac{1}{2}, \ \&c.$$

which are read two and one-fourth, three and two-thirds, &c.

Q. How many kinds of vulgar fractions? What are they? What is a proper fraction? Is its value greater or less than 1? Give an example of a proper fraction. What is an improper fraction? What is its value compared with 1? Give an example of an improper fraction. What is a simple fraction? What is a compound fraction? Give an example of a simple fraction. Of a compound fraction. What is a mixed number? Give an example of a mixed number? What kind of a fraction is $\tfrac{2}{3}$? $\tfrac{3}{4}$? $\tfrac{1}{4}$ of $\tfrac{3}{4}$? $\tfrac{1}{4}$ of 2? $\tfrac{1}{3}$ of $\tfrac{1}{2}$ of 4? $2\tfrac{1}{2}$? How is $2\tfrac{1}{2}$ read? $3\tfrac{1}{4}$? $5\tfrac{4}{7}$?

65. *If the numerator of a fraction be multiplied by a number, the denominator remaining the same, the value of the fraction will be increased as many times as there are units in the number.*

For since the numerator of a fraction expresses the number of the parts taken in the fraction (Art. **60**), if it be multiplied by a number, the number of parts taken will be increased as many times as there are units in the number, and the new fraction will be in like manner increased.

Thus, in the fraction $\tfrac{3}{2}$, the unit is divided into *halves*, and 3 *halves* are taken; but if the numerator 3 be multiplied by 2, the fraction becomes $\tfrac{6}{2}$, which is 2 times greater than it was before, since 6 halves are taken instead of 3; if the numerator be multiplied by 3, the fraction becomes $\tfrac{9}{2}$, which is 3 times greater than before, &c.

Q. What effect will there be in multiplying the numerator of a fraction by a number, the denominator remaining the same? How do you explain this? What does the numerator of a fraction express? What effect will be produced by multiplying it by a number? How is the unit divided in $\tfrac{3}{2}$? How many halves are taken? If you multiply 3 by 2, how many halves will be taken? Is $\tfrac{6}{2}$ greater than $\tfrac{3}{2}$? How many times greater? How many times is $\tfrac{9}{2}$ greater than $\tfrac{3}{2}$? How was $\tfrac{6}{2}$ obtained from $\tfrac{3}{2}$?

66. *If the numerator of a fraction be divided by a number, the denominator remaining the same, the value of the fraction will be diminished as many times as there are units in the number.*

For dividing the numerator is equivalent to diminishing the number of parts taken in the fraction, which will be as many times smaller as there are units in the number.

Thus, in the fraction $\frac{9}{2}$, 9 *halves* are taken, but if the numerator 9 be divided by 3, the fraction becomes $\frac{3}{2}$, in which but 3 *halves* are taken. $\frac{3}{2}$ is therefore 3 times smaller than $\frac{9}{2}$.

Q. If the numerator of a fraction be divided by a number, the denominator remaining the same, will the value of the fraction be changed, increased or diminished? Why diminished? If the numerator of the fraction $\frac{9}{2}$ be divided by 3, what will the resulting fraction be? Is $\frac{3}{2}$ greater or smaller than $\frac{9}{2}$? How many times smaller? Why?

67. *If the denominator of a fraction be multiplied by a number, the numerator remaining the same, the value of the fraction will be diminished as many times as there are units in the number.*

For the denominator expresses into what parts the unit is divided (Art. 60), and multiplying by a number will diminish the magnitude of these parts as many times as there are units in the number, and of course so many more of them will be required to make up the given unit.

Thus, in the fraction $\frac{1}{2}$, the unit is divided into *halves*, and multiplying the denominator by 2, the fraction becomes $\frac{1}{4}$, in which the unit is divided into *fourths*. But there are 2 *halves* in 1, and 4 *fourths* in 1; hence the fraction $\frac{1}{2}$ is 2 times greater than $\frac{1}{4}$.

Q. If the denominator of a fraction be multiplied by a number, the numerator remaining the same, will the fraction be increased or diminished? Why diminished? What does the denominator express? If the denominator be diminished, will the unit contain more or less of the parts expressed? How is the unit divided in $\frac{1}{2}$? In $\frac{1}{4}$? How many *halves* in 1? How many *fourths*? Is $\frac{1}{2}$ greater or less than $7\frac{1}{2}$? How many times less? Why?

68. *If the denominator of a fraction be divided by a number, the numerator remaining the same, the value of the fraction will be increased as many times as there are units in the number.*

For by dividing the denominator by a number, we increase the magnitude of the parts into which the unit is divided, and the number of the new parts required to make up the unit will be less in proportion as the parts have been increased.

Thus, in the fraction $\frac{1}{6}$ the unit is divided in *sixths*, and 6 sixths are required to make 1; but if the denominator 6 be divided by 3, the fraction becomes $\frac{1}{2}$, in which the unit is divided into *halves*; and since but 2 halves make 1, the fraction has been increased 3 times.

Q. If the denominator of a fraction be divided by a number, the numerator remaining the same, will the fraction be increased or diminished? Why increased? How is the unit divided in $\frac{1}{6}$? In $\frac{1}{2}$? How many sixths in 1? How many halves in 1? Is $\frac{1}{6}$ greater or less than $\frac{1}{2}$? Why less?

69. *The value of a fraction is not changed by multiplying its numerator and denominator by the same number.*

For when we multiply the numerator by this number, the value of the fraction is increased as many times as there are units in the number multiplied by (Art. 65), and when we multiply the denominator by the same number, the value of the fraction is in the same proportion diminished (Art. 67). Thus, $\frac{1}{2}$, $\frac{2}{4}$, $\frac{3}{6}$, $\frac{4}{8}$, &c., are equivalent fractions, since they are obtained by multiplying the terms of the fraction $\frac{1}{2}$ by 2, 3 and 4.

Q. Is the value of a fraction changed by multiplying its terms by the same number? Why not? Is $\frac{1}{2}$ greater or less than $\frac{2}{4}$?

70. *The value of a fraction is not changed by dividing its numerator and denominator by the same number.*

For, when we divide the numerator by a number, the value of the fraction is diminished as many times as there are units in the number (Art. 66); and when we divide the denominator by the same number, the value of the fraction is in like manner increased (Art. 68).

Thus $\frac{18}{20}$, $\frac{9}{10}$, $\frac{1}{5}$, are equivalent fractions.

Q. Is the value of a fraction changed by dividing its terms by a number? Why not?

REDUCTION OF VULGAR FRACTIONS.

71. Before the operations of Addition, Subtraction, Multiplication and Division can be performed upon Vulgar Fractions, it is frequently necessary to *reduce* them.

Reduction of Vulgar Fractions consists in changing the form of the fractions without altering their values.

Q. What is necessary to prepare Vulgar Fractions for the operations of addition, &c. ? In what does Reduction of Vulgar Fractions consist ?

CASE I.

72. *To reduce an improper fraction to an equivalent whole or mixed number—*

RULE.

Divide the numerator by the denominator.

EXAMPLES.

1. Reduce $\frac{28}{7}$ to its equivalent whole number.   7)28
Dividing 28 by 7, the quotient is 4.   ‾‾‾
    4

2. Reduce $\frac{29}{7}$ to an equivalent mixed number.   7)29
Dividing as before, 29 by 7, the quotient is 4,   ‾‾‾
and the remainder 1, which being placed over the $4\frac{1}{7}$
divisor 7, gives $\frac{1}{7}$ for the fractional part of the
mixed number.

It is evident that in neither of these cases is the value of the fraction changed, since by Art. 62 the value of every fraction was obtained by dividing the numerator by the denominator.

Q. How is an improper fraction reduced to an equivalent whole or mixed number ? Does the reduction alter the value of the fraction ? Why not ?

3. Reduce $\frac{97}{8}$ to an equivalent mixed number. *Ans.* $12\frac{1}{8}$.

4. Reduce $\frac{105}{9}$ to an equivalent mixed number. *Ans.* $11\frac{6}{9}$.

5. Reduce $\frac{29}{7}$, $\frac{108}{12}$, $\frac{114}{14}$, and $\frac{96}{8}$, to equivalent whole or mixed numbers.

CASE II.

73. *To reduce a mixed number to an equivalent improper fraction—*

RULE.

Multiply the whole number by the denominator of the fractional part of the mixed number, and to this product add the numerator. Place this result over the denominator.

EXAMPLES.

1. Reduce $4\frac{2}{3}$ to an equivalent improper fraction.

Multiplying 4 by 3, we get 12, and adding the numerator 2 to this product, we get 14, which being placed over the denominator 3, we have $\frac{14}{3}$ for the equivalent improper fraction.

OPERATION.

$$4\frac{2}{3}$$
$$3$$
$$\overline{\quad}$$
$$14 \cdot$$
$$\overline{\quad}$$
$$3 \; Ans.$$

The reason for this rule is evident, since multiplying the 4 by 3, brings the entire part of the mixed number to thirds (Art. 58), and there being $\frac{12}{3}$ in 4, there are $\frac{14}{3}$, that is, $\frac{12}{3}+\frac{2}{3}$ in $4\frac{2}{3}$.

Q. How is a mixed number reduced to an equivalent improper fraction? How many *thirds* in 4? How many thirds in $4\frac{2}{3}$?

2. Reduce $48\frac{2}{5}$ to an equivalent improper fraction.

Ans. $\frac{242}{5}$.

3. Reduce $527\frac{1}{5}$, $1207\frac{9}{10}$, $3004\frac{8}{100}$, to equivalent improper fractions.   *Ans.* $\frac{1585}{5}$, $\frac{12079}{10}$, $\frac{300408}{100}$.

4. Reduce $2704\frac{25}{150}$, $10047\frac{1003}{10000}$, $980702\frac{1}{175}$, to equivalent improper fractions.

CASE III.

74. *To reduce a Vulgar Fraction to its lowest terms.*

A fraction is reduced to its *lowest terms* when there is no number greater than 1 which will divide at the same time its numerator and denominator.

From this definition we deduce the following

RULE.

Divide the numerator and denominator of the fraction by any number that will divide them both without a remainder; divide these quotients in the same way until there is number greater than 1 that will exactly divide them.

1. Reduce $\frac{175}{70}$ to its lowest terms.

Dividing the two terms of the fraction by 5, it becomes $\frac{35}{14}$; the numerator and denominator of this

OPERATION.

$5)\frac{175}{70}=7)\frac{35}{14}=\frac{5}{2}.$

fraction being now divided by 7, we have $\frac{5}{2}$ for the *lowest
terms* of the given fraction, since no number greater than 1
will divide 5 and 2 at the same time without a remainder.

Q. When is a fraction reduced to its lowest terms? How is a fraction reduced to its lowest terms? What are the lowest terms of
$\frac{4}{8}$? Of $\frac{8}{10}$? Can $\frac{2}{3}$ be reduced to lower terms? Why not?

75. If we could find the *greatest* number that would
divide at the same time the two terms of the fraction, its reduction to its lowest terms could be at once effected by dividing by this number.

Any number which will exactly divide two or more numbers is called a *common* divisor.

The greatest number that will exactly divide two or more
numbers, is called the *greatest common divisor*.

Two numbers may have any number of *common* divisors,
but they can only have one *greatest* common divisor.

To find the greatest number that will exactly divide the
two terms of a fraction, reduces itself then, to ascertain the
greatest common divisor to these terms.

Take the fraction $\frac{72}{90}$.

Now it is evident that if 72
will divide 90 without a remainder, 72 will be the greatest common divisor, since it
goes once in itself and no
more. But 72 will not ex-
actly divide 90, since we have
a remainder 18. Now if 18

OPERATION.

$72)90(1$
$\quad 72$
$\quad\overline{}$
$18)72(4$
$\quad 72$
$\quad\overline{}$
$18)\frac{72}{90}=\frac{4}{5}$ lowest terms.

will divide 72, it will also divide 90, since 90 is equal to
$72+18$. But 18 does divide 72; it also divides 90: hence
18 is a *common* divisor to 72 and 90. But 18 is also the
greatest common divisor, since the greatest common divisor
must be contained at least once more in 90 than in 72. The
greatest common divisor cannot then be greater than the
difference between the numbers, which is in this case 18.

Hence 18 is the greatest common divisor sought, and dividing the two terms of the given fraction by it, we have $\frac{4}{5}$, which are the lowest terms of the fraction $\frac{72}{90}$.

76. Hence, to *reduce a fraction to its lowest terms by means of the greatest common divisor,* we have the following

RULE.

I. *Divide the greater term of the fraction by the less, and this divisor by the remainder, and so on, continuing to divide each divisor by the last remainder until nothing remains: the last divisor will be the greatest common divisor.*

II. *Divide each term of the fraction by the greatest common divisor.*

Q. If we knew the greatest number that would divide the two terms of a fraction, how might its reduction to its lowest terms be effected? What is a common divisor to two or more numbers? What is their greatest common divisor? How many common divisors may two or more numbers have? How many greatest common divisors? Have 2 and 3 a common divisor? Have 4 and 6? What is it? What is their greatest common divisor? Can the greatest common divisor to two numbers exceed the less of the two numbers? Why not? What is the rule for reducing a fraction to its lowest terms by means of the greatest common divisor? When the greatest common divisor is found, what is done with it?

EXAMPLES.

1. Reduce $\frac{63}{81}$ to its lowest terms by the greatest common divisor.

OPERATION.

$$63)81(1$$
$$63$$
$$\overline{}$$
$$18)63(3 \qquad 9)\frac{63}{81}=\frac{7}{9} \text{ lowest terms.}$$
$$54$$
$$\overline{}$$

greatest com. div.$=9)18(2$
$$18$$
$$\overline{}$$

Here the greater term 81 being divided by 63, we have a remainder 18, which is used as a divisor, and 63 for a dividend. We have by this second division a remainder 9, which we find exactly divides the last divisor; hence 9 is the greatest common divisor.

Dividing now 63 and 81 by 9, we have $\frac{7}{9}$ as the lowest terms of the fraction.

2. Reduce $\frac{172}{1118}$ to its lowest terms by the greatest common divisor. *Ans.* $\frac{2}{13}$.

3. Reduce $\frac{132}{144}$ to its lowest terms by the greatest common divisor. *Ans.* $\frac{11}{12}$.

4. Reduce $\frac{1049}{8392}$ to its lowest terms by the greatest common divisor. *Ans.* $\frac{1}{8}$

5. Reduce $\frac{1234}{11082}$, $\frac{105}{7885}$, $\frac{84}{1280}$, to their lowest terms by the greatest common divisor.

CASE IV.

77. *To reduce a compound fraction to its equivalent simple one —*

RULE.

Multiply the numerators together for a new numerator, and the denominators together for a new denominator.

EXAMPLE.

1. Reduce $\frac{3}{4}$ of $\frac{5}{7}$ to an equivalent simple fraction.

$$\frac{3 \times 5}{4 \times 7} = \frac{15}{28} \ Ans.$$

Here we multiply the numerators 3 and 5 together, and we get 15 for the new numerator; and then multiplying 4 and 7 together, we have 28 for the new denominator. $\frac{15}{28}$ is the simple fraction sought. For $\frac{1}{4}$ of $\frac{5}{7}$ is $\frac{5}{28}$, and 3 times $\frac{5}{28}$ is $\frac{15}{28}$.

When the compound fractions contain mixed numbers, they must be reduced to improper fractions as in Case II., and then the numerators and denominators multiplied together as just explained.

When the resulting simple fraction can be reduced to lower terms, the rule in Case III. should be applied.

2. Reduce $2\frac{1}{3}$ of $\frac{3}{4}$ to an equivalent improper fraction.

$2\frac{1}{3}$ reduced to an improper fraction is $\frac{7}{3}$.

$$\frac{7}{3} \text{ of } \frac{3}{4} = \frac{7 \times 3}{3 \times 4} = \frac{21}{12} \ Ans.$$

3. Reduce $\frac{3}{4}$ of $\frac{5}{8}$ of $\frac{1}{7}$ to an equivalent simple fraction. *Ans.* $\frac{15}{168} = \frac{5}{56}$.

4. Reduce $2\frac{1}{4}$ of $\frac{3}{5}$ of $4\frac{1}{2}$ to an equivalent simple fraction. *Ans.* $\frac{243}{40}$

D

5. Reduce $6\frac{1}{3}$ of $5\frac{1}{2}$ of 4 to an equivalent simple fraction.

Ans. $\frac{836}{6}=\frac{418}{3}$.

6. Reduce $\frac{2}{3}$ of 1 of $5\frac{1}{3}$ of $4\frac{1}{6}$ to an equivalent simple fraction. *Ans.* $\frac{1200}{38}=\frac{100}{3}$.

Q. How is a compound fraction reduced to an equivalent simple one? How do you proceed when the compound fraction contains mixed numbers? If the resulting simple fraction is not reduced to its lowest terms, what do you do?

CASE V.

78. *To reduce fractions having different denominators to equivalent fractions having the same or a common denominator—*

RULE.

I. *Reduce the whole or mixed numbers to equivalent proper or improper fractions.*

II. *Multiply each numerator by all the denominators except its own for a new numerator, and all the denominators together for a new denominator.*

EXAMPLES.

1. Reduce $\frac{1}{2}$, $\frac{3}{4}$, $\frac{4}{3}$ to fractions having the same or a common denominator.

$1 \times 4 \times 3 = 12$, numerator of the 1st fraction.
$3 \times 2 \times 3 = 18$, numerator of the 2d fraction.
$4 \times 2 \times 4 = 32$, numerator of the 3d fraction.
$2 \times 4 \times 3 = 24$, common denominator.

The new fractions therefore are

$$\frac{12}{24}, \frac{18}{24} \text{ and } \frac{32}{24}.$$

In this example the numerator 1 of the first fraction has been multiplied by the denominators 4 and 3 of the other fractions, to form the first new numerator; the numerator 3 of the second fractions by the denominators 2 and 3, to form the second new numerator; and the numerator 4 by the denominators 2 and 3, to form the third new numerator. The denominators were then multiplied together to form the new denominator.

It is evident that in this process the values of the fractions have not been altered, since the two terms of each fraction have been multiplied by the same number. (Art. 65.) *Those of the first fraction are multiplied by 4 and 3, those of the second by 2 and 3, and those of the third by 2 and 4.*

2. Reduce $2\frac{1}{4}$ and $\frac{1}{3}$ of $3\frac{1}{4}$ to fractions having a common denominator.

$$2\frac{1}{4}=\frac{9}{4}\text{ ; }\tfrac{1}{3}\text{ of }3\frac{1}{4}=\tfrac{1}{3}\text{ of }\frac{13}{4}=\frac{13}{8}.$$

$\frac{9}{4}$ and $\frac{13}{8}$ reduced to a common denominator become

$$\frac{72}{32}\text{ and }\frac{52}{32}.$$

Note.—The reduction of fractions to the same denominator may be often facilitated by multiplying the numerator and denominator of each fraction by such a number as will give the fractions the same denominator. Thus in the fractions $\frac{3}{7}$, $\frac{5}{28}$, $\frac{9}{14}$, if we multiply the numerators and denominators of the first fraction by 4, and those of the last by 2, the three fractions will have the same denominator, and will be

$$\frac{12}{28},\ \frac{5}{28},\ \frac{18}{28}.$$

Q. How are fractions having different denominators reduced to fractions having the same denominator? Does the operation alter the value of the fraction? Why not? How may the reduction be facilitated? By what would you multiply the terms of the fraction $\frac{1}{2}$ to make it have the same denominator as the fraction $\frac{3}{4}$?

3. Reduce $\frac{7}{8}$, $\frac{15}{6}$ and 37, to fractions having the same denominator. *Ans.* $\frac{525}{600}$, $\frac{1080}{600}$, $\frac{22200}{600}$.

4. Reduce $\frac{1}{2}$, $\frac{2}{3}$, $\frac{3}{4}$, $\frac{5}{8}$ and $\frac{7}{9}$, to fractions having the same denominator. *Ans.* $\frac{576}{1152}$, $\frac{768}{1152}$, $\frac{864}{1152}$, $\frac{720}{1152}$, $\frac{1344}{1152}$.

5. Reduce $1\frac{1}{8}$, $\frac{3}{4}$ of $1\frac{1}{2}$, $\frac{7}{12}$ and $\frac{5}{8}$ to fractions having the same denominator. *Ans.* $\frac{8448}{11520}$, $\frac{12960}{11520}$, $\frac{6720}{11520}$, $\frac{7200}{11520}$.

6. Reduce $4\frac{1}{8}$, $8\frac{1}{7}$ and $2\frac{1}{4}$ of 5 to fractions having the same denominator. *Ans.* $\frac{618}{128}$, $\frac{1026}{128}$, $\frac{1575}{128}$.

OF ADDITION OF VULGAR FRACTIONS.

79. Vulgar Fractions having been prepared by the reductions just explained for the operations of addition, subtraction, &c., we may now proceed to show how fractions are added.

Addition of fractions teaches us how to express by a single fraction the total value of several fractions.

Let it be required to add $\frac{1}{4}$ and $\frac{3}{4}$ together.

In both of these fractions the unit is divided into *halves* (Art. 60), and *one* of these halves is taken in the first fraction, and *three* of them in the second. There must therefore be *four* halves altogether, and the sum is $\frac{4}{2} = \frac{1+3}{2}$.

Again, add $\frac{3}{4}$ and $\frac{5}{4}$ together. Here the unit is divided into *fourths* in both cases, and 3 fourths are taken in the first fraction and 5 fourths in the second. There will then be 8 fourths altogether. Hence

$$\frac{3}{4} + \frac{5}{4} = \frac{3+5}{4} = \frac{8}{4} \; Ans.$$

We conclude, therefore, that for the addition of Vulgar Fractions, which have the same denominator, we have the following

RULE.

Add the numerators together for a new numerator, and set the sum over the common denominator. The result will be the sum required.

Q. For what operations does reduction of Vulgar Fractions prepare them ? What does addition of fractions teach ? How is the unit divided in the fraction $\frac{1}{2}$? In $\frac{3}{2}$? How many halves are taken in $\frac{1}{2}$? In $\frac{3}{2}$? How many halves in $\frac{1}{2}$ and $\frac{3}{2}$ together ? How many *fourths* in $\frac{3}{4}$? In $\frac{5}{4}$? In $\frac{3}{4}$ and $\frac{5}{4}$ together ? In $\frac{2}{4}$, $\frac{5}{4}$ and $\frac{8}{4}$? What is the rule for adding fractions which have a common denominator ?

EXAMPLES.

1. Add $\frac{2}{3}$, $\frac{5}{3}$, $\frac{8}{3}$ and $\frac{10}{3}$ together.

$$\frac{2+5+8+10}{3} = \frac{25}{3} = 8\frac{1}{3} \; Ans.$$

2. Add $\frac{1}{7}$, $\frac{3}{7}$, $\frac{5}{7}$ and $\frac{9}{7}$ together.   *Ans.* $\frac{18}{7}$.

3. Add $\frac{3}{8}$, $\frac{5}{8}$, $\frac{6}{8}$ and $\frac{9}{8}$ together.   *Ans.* $\frac{23}{8}$.

4. Add $\frac{1}{17}$, $\frac{4}{17}$, $\frac{7}{17}$ and $\frac{13}{17}$ together.   *Ans.* $\frac{25}{17}$.

80. Fractions are frequently presented for addition which have not a common denominator. Thus, let it be required to add $\frac{3}{4}$ and $\frac{1}{4}$ together. But by Case V. they can be brought to the same denominator, and the fractions are then added as has just been explained.

If any mixed number or compound fractions occur, they must be first brought to equivalent simple fractions, by the rules explained in Case II. and Case IV., and then add as in the above rule.

Q. What is necessary before you can add fractions which have not the same denominators? How is this done? After reducing them to the same denominator, how do you proceed? Suppose mixed numbers or compound fractions occur? After the reduction is effected, what do you do?

EXAMPLES.

1. Add $2\frac{1}{3}$ and $\frac{5}{6}$ of $\frac{1}{4}$ together.

$$2\frac{1}{3}=\frac{7}{3} \text{ and } \frac{5}{6} \text{ of } \frac{1}{4}=\frac{5\times1}{6\times4}=\frac{5}{24}.$$

Here the given fractions have been reduced to the two simple fractions $\frac{7}{3}$ and $\frac{5}{24}$. Multiplying the two terms of the first fraction by 8, it becomes $\frac{56}{24}$, and this fraction having the same denominator with $\frac{5}{24}$, we have

$$\frac{56}{24}+\frac{5}{24}=\frac{56+5}{24}=\frac{61}{24} \text{ Ans.}$$

2. Add 3, $2\frac{1}{4}$, $\frac{1}{3}$ of 2 and $\frac{3}{4}$ together.   *Ans.* $\frac{137}{20}=6\frac{17}{20}$.

3. Add $\frac{1}{7}$, $\frac{1}{3}$, $4\frac{3}{5}$ and $6\frac{1}{2}$ together.   *Ans.* $\frac{2309}{212}=10\frac{411}{212}$.

SUBTRACTION OF VULGAR FRACTIONS.

81. *Subtraction of fractions has for its object to find how much one fraction exceeds another.*

Let it be required to subtract $\frac{1}{2}$ from $\frac{3}{2}$. There being *one-half* in $\frac{1}{2}$, and three halves in $\frac{3}{2}$, the excess of $\frac{3}{2}$ over $\frac{1}{2}$ is two halves, or $\frac{2}{2}=\frac{3-1}{2}$.

Again, to subtract $\frac{3}{4}$ from $\frac{5}{4}$. The unit being in both cases divided into fourths, and five fourths being taken in the greater fraction, and three fourths in the less, the excess is *two* fourths, or $\frac{2}{4}=\frac{5-3}{4}$.

Hence, *to subtract fractions which have the same de minator, we have the following*

RULE.

Subtract the less numerator from the greater, and place the excess over the common denominator.

EXAMPLES.

1. From $\frac{8}{11}$ take $\frac{4}{11}$.   *Ans.* $\frac{4}{11}$.
2. From $\frac{12}{15}$ take $\frac{3}{15}$, and from the remainder take $\frac{4}{15}$.
                                                     Ans. $\frac{5}{15}$.
3. From $\frac{34}{113}$ take $\frac{21}{113}$, and then $\frac{6}{113}$.   *Ans.* $\frac{9}{113}$.

82. If the given fractions have not the same denominators, they must be reduced to the same denominator, as explained in Case V., and then subtracted as above.

Compound fractions and mixed numbers must be reduced to equivalent simple fractions as in Case II. and Case IV.

Q. What is the object of subtraction of fractions? How many halves in $\frac{3}{2}$? In $\frac{1}{2}$? How many more halves in $\frac{3}{2}$ than in $\frac{1}{2}$? How is the unit divided in $\frac{3}{4}$? In $\frac{5}{4}$? How many fourths in $\frac{3}{4}$? In $\frac{5}{4}$? How many more fourths in $\frac{5}{4}$ than in $\frac{3}{4}$? What is the rule for subtracting fractions which have a common denominator? If the fractions have not the same denominator, what do you do? How are they reduced to the same denominator? If the fractions be compound, what do you do? If mixed numbers?

EXAMPLES.

1. From $10\frac{3}{4}$ take $\frac{1}{2}$ of $\frac{4}{7}$.

$$10\frac{3}{4} = \frac{43}{4}, \quad \frac{1}{2} \text{ of } \frac{4}{7} = \frac{4}{14}.$$

Multiplying now the terms of the fraction $\frac{43}{4}$ by 7, and those of the fraction $\frac{4}{14}$ by 2 (Art. 74), to reduce them to the same denominator, they become $\frac{301}{28}$ and $\frac{8}{28}$.

$$\text{Then } \frac{301}{28} - \frac{8}{28} = \frac{301-8}{28} = \frac{293}{28} \text{ } Ans.$$

2. From $\frac{1}{7}$ of $5\frac{1}{2}$ take $\frac{1}{4}$ of $\frac{2}{3}$ of 2.   *Ans.* $\frac{85}{126}$.
3. From $9\frac{5}{8}$ take $4\frac{7}{8}$.   *Ans.* $4\frac{3}{4}$.
4. From $2\frac{1}{2}$ of 3 take $\frac{5}{8}$ of $\frac{1}{2}$.   *Ans.* $7\frac{1}{12}$.

MULTIPLICATION OF VULGAR FRACTIONS.

83. Multiplication of Vulgar Fractions has the same object as multiplication of whole numbers, and is a short way of doing addition.

In multiplication the multiplicand is to be repeated as many times as there are units in the multiplier. (Art. 26.)

Thus, to multiply $\frac{2}{3}$ by $\frac{4}{5}$ is the same as taking $\frac{4}{5}$ times the fraction $\frac{2}{3}$, that is, 4 times the fifth part of $\frac{2}{3}$. But in multiplying the denominator 3 by 5, we change the *thirds* into *fifteenths*, or into parts 5 times smaller (Art. 67). These parts are now to be taken 4 times, thus—

$$\tfrac{2}{15}+\tfrac{2}{15}+\tfrac{2}{15}+\tfrac{2}{15}=\frac{2\times4}{15}=\tfrac{8}{15}.$$

Hence, to multiply fractions together, we have the

RULE.

I. *Reduce the compound fractions or mixed numbers to equivalent simple fractions.*

II. *Multiply the numerators together for a new numerator, and the denominators together for a new denominator.*

Q. What is the object of the multiplication of fractions? In multiplication, how often is the multiplicand repeated? In multiplying $\frac{2}{3}$ by $\frac{4}{5}$, how many times is $\frac{2}{3}$ to be repeated? $\frac{4}{5}$ times is equivalent to what? How is the $\frac{1}{5}$ part of $\frac{2}{3}$ taken? When the denominator 3 is multiplied by 5, what is the denominator of the fraction? How now may $\frac{2}{15}$ be taken 4 times? What is the rule for multiplication of fractions?

EXAMPLES.

1. Multiply $3\frac{1}{4}$ by $\frac{1}{2}$ of $\frac{3}{9}$.

$$3\tfrac{1}{4}=\tfrac{13}{4},\quad \tfrac{1}{2}\text{ of }\tfrac{3}{9}=\tfrac{3}{18}.$$

The fractions are now reduced to the simple fractions $\frac{13}{4}$ and $\frac{3}{18}$, and may be multiplied thus—

$$\tfrac{13}{4}\times\tfrac{3}{18}=\frac{13\times3}{4\times18}=\tfrac{39}{72}\; Ans.$$

2. Multiply $\frac{1}{2}$ of $\frac{1}{4}$ of $\frac{2}{5}$ by $\frac{1}{3}$ of $\frac{9}{10}$ of $4\frac{1}{2}$.   Ans. $\frac{81}{400}$

3. Multiply $3\frac{2}{7}$ by $4\frac{11}{33}$.   Ans. $14\frac{11}{21}$.

4. *Multiply $\frac{1}{2}$ of $\frac{2}{3}$ by $\frac{1}{2}$ of $\frac{4}{5}$ of $\frac{5}{8}$.*   Ans. $\frac{1}{9}$.

84. Since a whole number may be expressed fraction-ally by writing 1 below it as a denominator (Art. 61), it follows that *to multiply a fraction by a whole number, or a whole number by a fraction, we have only to multiply the numerator of the fraction by the whole number, and set the product over the denominator of the fraction.*

Thus, to multiply $\frac{4}{5}$ by 7, since $7 = \frac{7}{1}$, we have

$$\tfrac{4}{5} \times 7 = \tfrac{4}{5} \times \tfrac{7}{1} = \frac{4 \times 7}{5} = \tfrac{28}{5}.$$

Q. How may a whole number be expressed fractionally? How may a fraction be multiplied by a whole number?

EXAMPLES.

1. Multiply $3\frac{4}{5}$ by 15.   *Ans.* 57.
2. Multiply $2\frac{1}{4}$ of $\frac{3}{4}$ by 9.   *Ans.* $1\frac{35}{8}$.
3. Multiply 12 by $\frac{3}{4}$ of $\frac{4}{5}$.   *Ans.* $7\frac{1}{5}$.
4. Multiply $2\frac{1}{4}$ of $\frac{4}{9}$ by $\frac{1}{2}$ of $\frac{4}{5}$, and then by 6.

                                  Ans. $\frac{112}{83}$.

DIVISION OF VULGAR FRACTIONS.

85. Division of Fractions, like that of whole numbers, is to find how often one fraction is contained in another.

Thus, to divide $\frac{4}{5}$ by $\frac{2}{3}$ is to find how many times $\frac{4}{5}$ contains $\frac{2}{3}$. But it is plain that $\frac{2}{3}$ will go in $\frac{4}{5}$ 3 times oftener than if the divisor were the whole number 2. Then, to divide $\frac{4}{5}$ by $\frac{2}{3}$ we must first divide by 2 and then multiply by 3, or what is the same thing, take 3 times the *half* of $\frac{4}{5}$, or $\frac{3}{2}$ of $\frac{4}{5} = \frac{12}{10}$.

Again, to divide $\frac{1}{2}$ by $\frac{5}{7}$. Here $\frac{5}{7}$ will go in $\frac{1}{2}$ *seven* times oftener than the whole number 5. Hence, the quotient will be 7 times the 5th of $\frac{1}{2}$, or $\frac{7}{5}$ of $\frac{1}{2} = \frac{7}{10}$. In both of these cases, it will be seen, that to obtain the quotient the terms of the divisor have been inverted: that is, the numerator become the denominator, and the denominator has be- ? *the numerator, and then the fractions were multiplied.* uce, *for the division of fractions, we have the following*

RULE.

I. *Reduce compound fractions and mixed numbers to their equivalent simple fractions.*

II. *Invert the terms of the divisor and then multiply the dividend by the divisor thus inverted.*

Q. What is division of fractions? To divide $\frac{4}{5}$ by $\frac{2}{3}$ is equivalent to what? How many more times will $\frac{2}{3}$ go in $\frac{4}{5}$ than the whole number 2? How then will you obtain the quotient of $\frac{4}{5}$ by $\frac{2}{3}$? Will $\frac{5}{5}$ go more times in $\frac{1}{5}$ than 5 will? How many more times? What change does the divisor of a fraction undergo in the process of finding the quotient? What is meant by inverting the terms of a fraction? Invert the terms of the fraction $\frac{2}{5}$, $\frac{4}{5}$. What is the rule for division of fractions?

86. To divide a fraction by a whole number, we put the whole number under the form of a fraction by writing 1 below it as a denominator (Art. 61), and then divide as in the rule.

Thus, to divide $\frac{3}{4}$ by 5. 5 expressed as a fraction is $\frac{5}{1}$. Then,

$$\frac{3}{4} \div \frac{5}{1} = \frac{3}{4} \times \frac{1}{5} = \frac{3}{4 \times 5} = \frac{3}{20} \; Ans.$$

Hence, to divide a fraction by a whole number, we multiply the denominator of the fraction by the whole number, and place the product under the numerator of the fraction.

Q. How is a fraction divided by a whole number? How do you express 5 as a fraction?

87. To divide a whole number by a fraction, we put the whole number under a fractional form, as in the last article, and then divide as the rule directs.

Thus, to divide 2 by $\frac{5}{3}$. 2 written fractionally is $\frac{2}{1}$. Then

$$\frac{2}{1} \div \frac{5}{3} = \frac{2}{1} \times \frac{3}{5} = \frac{2 \times 3}{5} = \frac{6}{5} \; Ans.$$

Hence, to divide a whole number by a fraction, we multiply the whole number by the denominator of the fraction, and place the product over the numerator of the fraction.

Q. How may a whole number be divided by a fraction?

EXAMPLES.

1. Divide $12\frac{3}{4}$ by $6\frac{2}{3}$.	*Ans.* $1\frac{73}{80}$.
2. Divide $4\frac{7}{11}$ by $15\frac{5}{8}$.	*Ans.* $\frac{408}{1375}$.
3. Divide $19\frac{1}{2}$ by 9.	*Ans.* $2\frac{1}{6}$.

D 2

4. Divide 18 by $5\frac{2}{3}$.   *Ans.* $3\frac{3}{17}$.

5. Divide $\frac{4}{5}$ of $\frac{6}{8}$ by $7\frac{1}{2}$.   *Ans.* $\frac{4}{75}$.

6. Divide $\frac{2}{3}$ of $3\frac{1}{4}$ of 5 by $\frac{4}{8}$ of $\frac{4}{7}$.   *Ans.* $85\frac{5}{18}$.

APPLICATIONS UPON VULGAR FRACTIONS.

1. The circumference of the earth is about 25000 miles; how long would it require an individual to walk around it, supposing that he travels $25\frac{2}{4}$ miles per day. *Ans.*

Q. How many miles round the earth?

2. By the Constitution of the United States, representatives in Congress are apportioned among the several States according to their respective numbers, which are determined by adding to the whole number of free persons $\frac{3}{5}$ of the slaves. How many representatives is each of the following States entitled to, the number of free persons and slaves in each being in 1840 as stated, and *one* representative being allowed for every 70,680 of the population represented?

States.	Free Persons.		Slaves.	Population represented.		Number of Members.
Delaware..	$75,480 + \frac{3}{5}$	of	$2,605 =$	385215	$\div 70,680 =$	$1\frac{6363}{10088}$
Maryland.	$380,282 + \frac{3}{5}$	of	$89,737 =$	$217 \frac{0}{5} 621$	$\div 70,680 =$	$6\frac{50321}{353400}$
Virginia..	$790,810 + \frac{3}{5}$	of	$448,987 =$	$530\frac{1}{5} 611$	$\div 70,680 =$	$15\frac{11}{353400}$
N. Carolina	$507,602 + \frac{3}{5}$	of	$245,817 =$		$\div 70,608 =$	
S. Carolina	$267,360 + \frac{3}{5}$	of	$327,038 =$		$\div 70,680 =$	
Georgia...	$410,448 + \frac{3}{5}$	of	$280,944 =$		$\div 70,680 =$	
Alabama..	$337,224 + \frac{3}{5}$	of	$253,532 =$		$\div 70,680 =$	
Mississippi	$180,440 + \frac{3}{5}$	of	$195,211 =$		$\div 70,680 =$	
Louisiana.	$183,959 + \frac{3}{5}$	of	$168,452 =$		$\div 70,680 =$	
Tennessee.	$646,151 + \frac{3}{5}$	of	$183,059 =$		$\div 70,680 =$	
Kentucky .	$597,570 + \frac{3}{5}$	of	$182,258 =$		$\div 70,680 =$	
Missouri ..	$325,462 + \frac{3}{5}$	of	$58,240 =$		$\div 70,680 =$	
Arkansas..	$77,639 + \frac{3}{5}$	of	$19,935 =$		$\div 70,680 =$	

Note.—The large fractions in the States of South Carolina, Alabama and Tennessee, are represented by the law of Congress fixing the ratio of representation, which gives to these States *one* more member than is shown in the column of answers.

Q. How is the population in the several States estimated in fixing *the numbers to be* represented in Congress? By what instrument is this principle fixed ? To whom is the power given to fix the ratio of representation ? What was the ratio fixed in 1841 ? In what States are the large fractions represented ? Which State has the most slaves ? Are the slaves numbered as free persons in fixing the number of members of Congress ? How are slaves enumerated ?

3. By a law of Congress the $\frac{1}{36}$ part of all the public lands in the States and Territories, is reserved for common schools in said States and Territories. What is the total number of acres thus reserved, and the total value of the same, estimating the land at $1\frac{1}{4}$ dollars per acre, the total number of acres of public lands in each State and Territory being as follows :

States.	Acres.	Acres for Common Schools.	Dolls.	Total value in Dollars.
Ohio	16,555,952	$\frac{16,555,952}{36} \times 1\frac{1}{4}$	=	Ans. 574,859$\frac{4}{9}$
Indiana	20,457,393	$\frac{20,457,393}{36} \times 1\frac{1}{4}$	=	Ans. 710,326$\frac{7}{43}$
Illinois	31,933,736	$\frac{31,933,736}{36} \times 1\frac{1}{4}$	=	Ans.
Missouri	40,241,436	$\frac{40,241,436}{36} \times 1\frac{1}{4}$	=	Ans.
Alabama	31,699,470	$\frac{31,699,470}{36} \times 1\frac{1}{4}$	=	Ans.
Mississippi	21,920,786	$\frac{21,920,786}{36} \times 1\frac{1}{4}$	=	Ans.
Louisiana	20,437,559	$\frac{20,437,559}{36} \times 1\frac{1}{4}$	=	Ans.
Michigan	31,118,392	$\frac{31,118,392}{36} \times 1\frac{1}{4}$	=	Ans.
Arkansas	31,468,911	$\frac{31,468,911}{36} \times 1\frac{1}{4}$	=	Ans.
Florida	36,755,840	$\frac{36,755,840}{36} \times 1\frac{1}{4}$	=	Ans.
Wisconsin	29,863,925	$\frac{29,863,925}{36} \times 1\frac{1}{4}$	=	Ans.
Iowa	7,082,832	$\frac{7,082,832}{36} \times 1\frac{1}{4}$	=	Ans.

Note.—The lands ceded by the Chickasaw Indians in Alabama and Mississippi, to be sold for their benefit, and the lands in Ohio and Indiana set apart by the deeds of cession of Virginia and Connecticut, and those sold for the benefit of the Indians, are excluded in this table.

Q. In what States and Territories are the United States public lands situated ? What amount is set apart for common schools ?

4. The usual proportion for making common mortar is to take *one-third* lime and *two-thirds* sand. How much lime and how much sand in 243$\frac{3}{4}$ bushels of mortar ?

Ans. 81$\frac{1}{4}$ bushels of lime.
162$\frac{1}{2}$ bushels of sand.

Q. Of what is common mortar composed ? What proportion of lime ? What of sand ?

5. The circumference of a circle is equal to its diameter multiplied by $\frac{355}{113}$. What is the circumference of the earth, its diameter being about 7957$\frac{3}{4}$ miles ? *Ans.*

Q. How may you obtain the circumference of a circle when you know its diameter ? What is the length of the earth's diameter ? What is its circumference ?

6. It is computed that a man travels on foot and on level ground about $31\frac{1}{4}$ miles per day. How many miles at this rate would a man travel in a year which contains $365\frac{1}{4}$ days ?   *Ans.*

Q. What is the computed rate of travel of a man on foot ? How many days and fractions of a day in a year ?

7. There are 544,743 white persons above 20 years of age, in the 26 States, who cannot read and write. What would be the cost of $3\frac{1}{2}$ years' schooling of the whole number, at $14\frac{3}{4}$ dollars for each person per year ?   *Ans.*

Q. How many white persons in the 26 States who cannot read and write ?

OF DECIMAL FRACTIONS.

88. When the unit 1 is divided into *tenths, hundredths, thousandths,* &c., the resulting fractions are called *Decimal Fractions.*

Thus, $\frac{1}{10}$, $\frac{4}{10}$, $\frac{1}{100}$, $\frac{12}{100}$, $\frac{1}{1000}$, $\frac{15}{1000}$, &c., are decimal fractions.

The denominator of a decimal fraction is not usually expressed, and to distinguish the numerator, which is alone written, from a whole number, a point, . called the *decimal point,* is placed on its left.

Thus, $\frac{2}{10}$ is written .2

$\frac{5}{10}$ "   .5

$\frac{25}{100}$ "   .25

$\frac{105}{1000}$ "   .105, &c. &c.

When the denominator of decimal fractions is expressed, *it is always 1, with as many ciphers to the right as there are figures in the numerator.*

Q. What are decimal fractions ? Are they usually expressed like other fractions ? Is the denominator usually expressed ? How is the numerator distinguished from a whole number ? What is the point called ? Where is it placed ? How would you express $\frac{3}{10}$ decimally ? $\frac{5}{10}$? Is $\frac{3}{4}$ a decimal fraction ? Why not ? Into how many equal parts is the unit divided in $\frac{1}{4}$? Is $\frac{7}{16}$ a decimal fraction? Why? How is the unit divided in it? How is it expressed decimally? What is the denominator of a decimal fraction?

89. The place next to the decimal point is called the *tenths place;* the next to the right the *hundredths place;* the next the *thousandths,* &c.; so that the same number decreases in a *tenfold* proportion as we proceed from-the decimal point to the right.

Thus, $\frac{4}{10}$ is written .4, the 4 being in the *tenths* place, and is 4 *tenths.*

$\frac{4}{100}$ is written .04, the 4 being in the hundredths place, and is 4 *hundredths.*

$\frac{4}{1000}$ is written .004, the 4 being in the thousandths place, and is 4 *thousandths.*

$\frac{4}{10000}$ is written .0004, the four being in the ten thousandths place, and is 4 *ten thousandths.*

Q. What is the place next to the decimal point called? The next? The next? How does the value of the same number vary as we proceed to the right in decimals? How is $\frac{4}{10}$ written decimally? What place does the 4 occupy? What is .4? How is $\frac{4}{100}$ written decimally? What place does the 4 take? How is $\frac{4}{1000}$ written decimally? What place does the 4 occupy? Is .4 greater or less than .04? How many times greater? Is .04 greater or less than .004? How many times greater? How many times is .004 less than .4?

90. Decimals are numerated from the decimal point to the right. Thus:

DECIMAL TABLE.

Tenths.	Hundredths.	Thousandths.	Tens of thousandths.	Hundreds of thousandths.	Millionths.
.4	7	6	9	8	5

The first figure is read four *tenths.*
The first two, forty-seven *hundredths.*

The first three, four hundred and seventy-six *thousandths*.

The first four, four thousand seven hundred and sixty-nine *ten thousandths*.

The first five, forty-seven thousand six hundred and ninety-eight *hundred thousandths*.

The whole number is four hundred and seventy-six thousand nine hundred and eighty-five *millionths*.

A *mixed* number is a whole number and a decimal, and as whole numbers decrease from the left to the right (Art. 8), as has just been shown to be the case with decimal fractions, it follows that a *mixed number may be written decimally by placing the decimal part to the right of the whole number, with the decimal point between them.*

Thus, *four* and *seven tenths* is written · 4.7

four and *seven hundreths* " 4.07

four and *seven thousandths* " 4.007

Q. How are decimals numerated? Read the decimal table? The first figure is what? The two first? The whole number? Read the fraction .014? .1004? .50075? .170403? Do decimals decrease or increase from the left to the right? How is it with whole numbers? What is a mixed number? How may a mixed number be written decimally? Write five and four ten thousandths. Seventeen and three millionths.

91. Since $\frac{5}{10}$, $\frac{50}{100}$, $\frac{500}{1000}$, $\frac{5000}{10000}$, &c., are fractions of equal value (Art. 69), their equivalent decimals, .5, .50, .500, .5000, &c., must also be of equal value. Hence, *placing ciphers on the right of a decimal fraction, does not alter its value.*

Again, $\frac{5}{10}$ is ten times greater than $\frac{5}{100}$, one hundred times greater than $\frac{5}{1000}$, and one thousand times greater than $\frac{5}{10000}$ (Art. 67); their equivalent decimals must vary in the same proportion. That is,

.5 is ten times greater than .05.

.5 is one hundred times greater than .005.

.5 is one thousand times greater than .0005, &c.

Hence, *placing a cipher on the left of a decimal fraction diminishes its value tenfold.*

.0005 is ten times less than .005, one hundred times less than .05, and one thousand times less than .5.

Q. What effect has placing ciphers on the right of a decimal? Why is not its value changed? Is $\frac{5}{10}$ of the same value as $\frac{50}{100}$? Why? How do .5 and .50 compare in value? If a cipher be placed on the left of a decimal, is the value of the fraction changed? Why? Is $\frac{5}{10}$ less or greater than $\frac{5}{100}$? Why greater? How much greater? What are their equivalent decimals? How do .5 and .05 compare in value? How do $\frac{5}{10}$ and $\frac{5}{1000}$ compare in value? What are their equivalent decimals? How do .5 and .005 compare in value? If one cipher be placed on the left of a decimal, how many times is its value diminished? If two? If three? If four? How do the following decimals compare with each other: .1, .01, .001, .0001, .00001? How do the following decimals compare with each other in value: .1, .10, .100, .1000, .10000?

OF ADDITION OF DECIMAL FRACTIONS.

92. Addition of Decimal Fractions is as readily performed as addition of whole numbers. The only difficulty consists in placing the decimal point.

Since in whole numbers units are placed under units, tens under tens, &c. (Art. 17), so *in addition of decimal fractions, tenths are placed under tenths, hundredths under hundredths, thousandths under thousandths,* &c., and the sum is then found as in whole numbers.

Thus, add 12.04, 5.307 and .0765 together.

In this example the tenths are placed under tenths, hundredths under hundredths, &c., which brings the decimal points directly under each other, and we then add as in whole numbers. The sum will evidently contain *tenths* only, if only tenths are to be added; it will

OPERATION.
```
12.04
 5.307
  .0765
-------
17.4235
-------
```

contain *tenths* and *hundredths*, if tenths and hundredths are to be added, &c. The example taken containing tenths, hundredths, thousandths and tens of thousandths, the sum must contain them also. We have then the following

RULE.

I. *Place the numbers one under the other so that tenths fall under tenths, hundredths under hundredths,* &c.

II. *Add as in addition of simple numbers, pointing off as many decimal places from the right hand as there are in the number which has the greatest number of de places among the given numbers.*

Q. What is the only difficulty in addition of decimals? How are the decimals written down? How many decimal places in the sum? What is the rule for adding decimals?

EXAMPLES.

1. Add 340.10, 1.004, .3, .0047, together.

Ans. 341.4087.

2. Add 7.10405, 30.04, .7632 and 104.300768 together.

Ans. 142.208018.

3. Add 3 tenths, 3 hundredths, 33 thousandths, 333 ten thousandths together.   *Ans.* .3963.

4. Add 4 tenths, 44 hundredths, 444 thousandths, 44 ten thousandths, and 4 millionths.   *Ans.* 1.288404.

5. Add seventeen and six hundredths, thirty-seven thousand four hundred and ninety-one hundred thousandths, 129 and 3 hundredths.   *Ans.* 146.46491.

SUBTRACTION OF DECIMAL FRACTIONS.

93. In subtraction of Decimal Fractions we have to find the difference between two decimal numbers. It differs but little from subtraction of whole numbers.

Let it be required to take 12.379 from 25.25.

Placing units under units, and tens under tens in the entire part of the two numbers, and tenths under tenths, hundredths under hundredths, &c., in the decimal part, we find there are no *thousandths* in the minuend under which to place the 9 *thousandths* of the subtrahend. But as placing a cipher on the right of a decimal does not change its value

OPERATION.

25.250
12.379
———
12.871
———

(Art. 91), we place a cipher in the thousandths place of the minuend, and then subtracting as in whole numbers, we have for the remainder 12.871. We have the following

RULE.

I. *Set the less number under the greater, placing tenths under tenths, hundredths under hundredths, &c., and when the number of decimal places is unequal in the two num-*

bers, make them equal by placing ciphers on the right of the number which has the smallest number of decimal places.

II. *Subtract as in whole numbers, pointing off in the remainder as many decimal places as there are in either of the given numbers.*

Q. What is subtraction of decimals? How are the numbers set down? If one of them has fewer number of decimal places, how may they be made equal? Does this alter the value of the decimals? Why not? How many decimal places are there in the remainder? What is the rule?

EXAMPLES.

1. From 187.125 take 39.6178.   *Ans.* 147.5072.
2. From 219.0769 take 137.187.   *Ans.* 81.8899.
3. From .631582 take .51907.   *Ans.* 112512.
4. From .525 take .02.   *Ans.* .505.
5. From 5 hundredths take 5 millionths. *Ans.* .049995.
6. From 6 and 1 thousandth take 1 and 24 millionths.

                                        Ans. 5.000976.

MULTIPLICATION OF DECIMAL FRACTIONS.

94. Let it be required to multiply .41 by .3.

These decimals expressed in the form of vulgar fractions are (Art. 88),

$$\tfrac{41}{100} \text{ and } \tfrac{3}{10}.$$

Which fractions, multiplied together, give us (Art. 83),

$$\frac{41 \times 3}{100 \times 10} = \tfrac{123}{1000} = .123 \text{ } Ans. \text{ (Art. 88.)}$$

Again, multiply .4 by .002.

$$.4 = \tfrac{4}{10} \text{ and } .002 = \tfrac{2}{1000}.$$

Hence, $.4 \times .002 = \tfrac{4}{10} \times \tfrac{2}{1000} = \tfrac{8}{10000} = .0008.$

As the product is 8 *ten thousandths*, it is necessary to place three ciphers on the left of the numerator 88, when the fraction is written decimally.

From these explanations we deduce the following

RULE.

I. *Multiply as in whole numbers, pointing off from the right hand of the product as many decimal places as there are in the multiplicand and multiplier together.*

II. *Should there not be as many figures in the product as are required, place ciphers on the left to supply the deficiency.*

Q. What is the rule for multiplication of decimals? How many decimal places do you point off in the product? If the number of figures in the product be not equal to the number of decimal places to be pointed off, how do you supply the deficiency?

EXAMPLES.

1. Multiply .4 by .3.   *Ans.* .12.

2. Multiply .4 by .03.   *Ans.* .012.

3. Multiply .4 by .003.   *Ans.* .0012.

4. Multiply .04 by .0003.   *Ans.* .000012.

5. Multiply .1004 by .075.   *Ans.* .0075300.

6. Multiply 10.01 by 5.105.   *Ans.* 51.10105.

7. Multiply 8 and 2 thousandths by 4 millionths.
                                     Ans. .000032008.

8. Multiply 325.05 by .0004.   *Ans.* .130020.

DIVISION OF DECIMAL FRACTIONS.

95. In division of whole numbers the dividend is equal to the product of the divisor by the quotient. It is the same in Decimal Fractions. Hence, the dividend will contain as many decimal places as there are in the divisor and quotient together (Art. 94). From which we conclude, *that the number of decimal places in the quotient will be equal to the excess of the number of decimal places in the dividend over that in the divisor.*

1. Divide 180 by 3.75.

In this example there being no decimal places in the dividend, two ciphers are placed on the right, which do not *change the value* of the dividend (Art. 91), and then dividing, the quotient is 48.

OPERATION.

$$3.75) 180.00 (48 \text{ } Ans.$$
$$\underline{1500}$$
$$3000$$
$$\underline{3000}$$
$$\cdots\cdots$$

Now, since the number of decimal places in the dividend is *equal* to that in the divisor, there will be no decimal places to point off in the quotient. The quotient is therefore a whole number.

2. Divide 2.3421 by 2.11.

OPERATION.

2.11) 2.3421 (1.11 (*Ans.*
211

232
211

211
211

. . .

In this example there being four decimal places in the dividend, and but two in the divisor, the quotient must also have two, which is therefore 1.11.

3. Divide .22875 by 2.5.

OPERATION.

2.5) .22875 (.0915 *Ans.*
225

37
25

125
125

. . .

In this example the dividend contains five places of decimals, and the divisor one: hence the quotient must contain *four*. But there being but three figures in the quotient, we place a cipher on the left, and the correct quotient is .0915.

Since the product in multiplication of decimals contains as many decimal places as there are in the multiplicand and multiplier together, the proof shows that .0915 is the true quotient.

PROOF.

.0915
2.5

4575
1830

.22875

From these examples we have for Division of Decimal Fractions the following

RULE.

I. *Divide as in whole numbers, pointing off from the right of the quotient as many decimal places as is equal to the excess of the number of decimal places in the dividend over that of the divisor.*

II. *Should the quotient not contain as many figures as is equal to this excess, place ciphers on the left to supply the deficiency.*

III. *Should the number of decimal places in the divisor exceed that in the dividend, place as many ciphers on the right of the dividend as shall make the whole number of decimal places in the dividend equal to that of the divisor. The quotient will then be an entire quantity.*

Q. What is the dividend equal to? How many decimal places must the dividend contain? How many decimal places must the quotient contain? If the dividend contain four decimal places and the divisor two, how many will there be in the quotient? What do you do if the number of figures in the quotient is not equal to the excess of the decimal places in the dividend over the divisor? What do you do when the divisor contains more decimal places than the dividend? Does placing ciphers on the right of the dividend change its value? What is the rule for division of decimal fractions?

96. When there is a remainder after all the figures in the dividend have been divided, we may continue the division by placing ciphers on the right of the dividend, regarding them as constituting a part of the decimals in the dividend.

Thus, to divide 1 by .3, we annex 4 ciphers, regarding them as decimals. As each operation gives the same remainder, the division may be continued indefinitely. The quotient being incomplete, it is written 3.333+. When one or more figures are continually repeated, the decimal is called a *recurring decimal.* The quotient 3.3333 is a *recurring decimal.*

$$.3\,)1.0000\,(\,3.333+$$

```
  9
 __
 10
  9
 __
 10
  9
 __
 10
  9
 __
  1, &c.
```

Q. How may the division be continued when there is a remainder? When the same remainder continually occurs, what do you conclude? How do you indicate that the quotient is incomplete? What is a recurring decimal?

EXAMPLES.

1. Divide .58875 by .75.   *Ans.* .785.

2. *Divide 182.5 by 2.5.*   *Ans.* 73.

3. *Divide 476 by .85.*   Ans. 560.

4. Divide 12.75 by 3.75. *Ans.* 3.4,
5. Divide 243.2 by 38. *Ans.* 6.4.
6. Divide 29 by .8. *Ans.* 36.25.
7. Divide .0024 by .018. *Ans.* .1333+.
8. Divide 8 by .11. *Ans.* .7272+.
9. Divide 76.75 by 3.25. *Ans.* 23.615+.
10. Divide .010001 by .01. *Ans.* 1.0001.
11. Divide 1 tenth by 1 millionth. *Ans.* 100000.
12. Divide 25 hundredths by 33 ten thousandths,
Ans. 75.7575+.

Place the decimal points correctly in the quotients of the following examples:

13. Divide 320.95 by 3.5. *Ans.*
14. Divide 320.095 by 3.5. *Ans.*
15. Divide .32095 by 3.5. *Ans.*
16. Divide .32095 by .35. *Ans.*
17. Divide 3209.5 by 35. *Ans.*
18. Divide 320.95 by 35. *Ans.*
19. Divide 32.095 by 35. *Ans.*
20. Divide 3.2095 by 35. *Ans.*
21. Divide .32095 by 35. *Ans.*
22. Divide .032095 by 35. *Ans.*
23. Divide 32095 by 3.5. *Ans.*
24. Divide 32095 by .35. *Ans.*
25. Divide 32095 by .035. *Ans.*
26. Divide 32095 by .0035. *Ans.*

97. We have seen (Art. 62) that a fraction denotes division, and that its value is found by dividing the numerator by the denominator. Thus, ½ is equal to the quotient of 1 by 2, but as 1 is less than 2, we annex a cipher to make the division possible, and then divide as follows: 2 in 1 we cannot, but since 1 unit is equal to 10 tenths, we can divide 10 tenths by 2, and the quotient is 5 tenths, or .5; .5 is then the value of the given fraction expressed de

OPERATION.

2) 1.0
———
.5

Again, $\frac{3}{4}$ is equal to 3 divided by 4. Annexing ciphers as before, we find that 4 in 3 we cannot, but since 3 units are equal to 30 tenths, 4 in 30 tenths goes 7 tenths times and 7 tenths over, but 2 tenths are equal to 20 hundredths, and 4 in 20 hundredths goes 5 hundredths times and none over.

OPERATION.

4) 3.00
———
.75

7 tenths and 5 hundredths, or 75 hundredths, is the value of the given fraction expressed decimally.

We conclude, therefore, *that to express the value of any vulgar fraction decimally, we place ciphers on the right of the numerator, regarding them as decimals, and then divide by the numerator, pointing off as many decimal places from the right of the quotient as there are ciphers annexed. When there are not as many figures in the quotient as there are ciphers used, ciphers must be prefixed to make up the deficiency.*

EXAMPLES.

1. Express the fraction $\frac{1}{4}$ decimally.   *Ans.* .25,

2. Reduce $\frac{1}{3}$ and $\frac{1}{6}$ to decimals.

Ans. .333+ and .1666+.

3. Reduce $\frac{11}{12}$ to the form of a decimal.. *Ans.* .9166+.

4. Express $\frac{12}{11}$ decimally.   *Ans.* 1.0909+.

5. Express $\frac{1}{6}^0$ decimally.   *Ans.* 16.666+.

6. Express $\frac{1}{7}^0$ decimally.   *Ans.* 14.285+.

Q. What does a fraction denote? How is its value found? How may a vulgar fraction be expressed decimally? How many tenths are there in 1 unit? How many hundredths? How many hundreds in 1 tenth? How many decimals are pointed off in the quotient? If the number of figures in the quotient is not equal to the number of decimals to be pointed off, how do you supply the deficiency?

APPLICATIONS UPON DECIMAL FRACTIONS.

1. Find the value of the Flour shipped from the United States in each of the years from 1790 to 1838; the number of barrels shipped each year, and the average prices in Philadelphia being as follows:

Y'rs.	No. of Barrels.	Average price. Dollars.	Total value. Dollars.
1790	724,623	× 5.56	= 4,028,903.88
1791	619,681	× 5.22	= 3,234,734.82
1792	824,464	× 5.25	= 4,328,436.00
1793	1,074,639	× 5.90	= 6,340,370.10
1794	846,010	× 6.90	
1795	687,369	× 10.60	
1796	725,194	× 12.50	
1797	515,633	× 8.91	
1798	567,558	× 8.20	
1799	519,265	× 9.66	
1800	653,052	× 9.86	
1801	1,102,444	× 10.40	
1802	1,156,248	× 6.90	
1803	1,311,853	× 6.73	
1804	810,008	× 8.23	
1805	777,513	× 9.70	
1806	782,724	× 7.30	
1807	1,249,819	× 7.17	
1808	263,813	× 5.69	
1809	846,247	× 6.91	
1810	798,413	× 9.37	
1811	1,445,012	× 9.95	
1812	1,443,492	× 9.83	
1813	1,260,942	× 8.92	
1814	193,274	× 8.60	
1815	862,739	× 8.71	
1816	729,053	× 9.78	
1817	1,479,198	× 11.69	
1818	1,157,697	× 9.96	
1819	750,660	× 7.11	
1820	1,177,036	× 4.72	
1821	1,056,119	× 4.78	
1822	827,865	× 6.58	
1823	756,702	× 6.82	
1824	996,792	× 5.68	
1825	813,906	× 5.10	
1826	857,820	× 4.65	
1827	868,496	× 5.23	
1828	860,809	× 5.60	
1829	837,385	× 6.33	
1830	1,227,434	× 4.83	
1831	1,806,529	× 5.67	
1832	864,919	× 5.72	
1833	955,766	× 5.63	
1834	835,352	× 5.17	
1835	779,396	× 5.88	
1836	505,400	× 7.99	
1837	319,719	× 9.37	
1838	448,161	× 7.79	

OF COMPOUND NUMBERS.

98. In all the operations which we have explained, we have considered the numbers in each example as expressing entire or fractional parts of the same kind of unit; that is, all dollars or parts of dollars; all yards or parts of yards, &c. To those numbers which are composed of the same kind of unit we have given the name of *Simple Numbers*.

Compound Numbers are those which are composed of units of different kinds.

Thus, 8 dollars 5 cents and 3 mills is a compound number.
10 pounds 14 shillings and 4 pence is a compound number.

Q. What kind of numbers are those upon which we have been operating? What name was given to these numbers? What are compound numbers? Give some examples of compound numbers.

99. Before explaining the operations used in compound numbers, we will first show the different kinds of units.

In Art. 2 we have defined a *unit to be a single quantity which is used to compare quantities of the same kind with each other.*

Thus, were I to ask Mary how much more ribbon she had in her work-basket than Ann had, she might apply the middle finger of her right hand to her ribbon as often as it would contain it, and say, her ribbon was 4 *fingers long;* and by applying the same finger to Ann's ribbon in the same way, she might find that there were 2 *fingers* in Ann's ribbon. Mary has therefore 2 *fingers* more ribbon than Ann has. In this example the middle finger of Mary's right hand was the *unit* by which the quantity of ribbon in the two cases was compared.

Again, were I to ask John how much longer his top-cord was than Henry's, he might take his pencil, and applying it to both cords, find that his cord contained the pencil 6½ times, and Henry's contained it 5 times. John's cord would be 1½ times the pencil longer than Henry's. Here the *pencil is the unit* of measure.

Again, find how many more chestnuts James has in his bag than William has in his. You might take a tin cup, and on measuring find that James's filled the cup 20 times, and William's 15 times. James has 5 cups more of chestnuts than William, and the *tin cup* was the *unit* of measure.

What has been done in these three cases might be extended to any kinds of measure whatever.

Q. What has a unit been defined to be? How do you explain this by means of the ribbon? What is the unit in this case? In the case of the top-cords what is the unit? In the chestnuts?

100. But as the middle fingers of all little girls' hands are not of the same length, and as measures for pencils and cups also vary, it is found necessary to fix by *law* the magnitude of the different units used, in order that the measures may be *uniform*.

Further, as convenience would require a small unit for small quantities measured, and a larger unit for larger quantities, different units are used, whose magnitude corresponds to the quantities measured.

The following tables comprise the divisions of the different kinds of units used in the United States. They should be carefully committed to memory.

Q. Why will not the middle finger answer as a unit? How are the units of measure made uniform? What arrangement is made to suit the quantity measured?

101. UNITED STATES CURRENCY, OR FEDERAL MONEY.

		E.	$	d.	cts.	mills.
10 Mills make 1 Cent.	*Sign* Ct.	1	=10	=100	=1000	=10000
10 Cents " 1 Dime.	*Sign* d.		1	= 10	= 100	= 1000
10 Dimes " 1 Dollar.	*Sign* $			1	= 10	= 100
10 Dollars " 1 Eagle.	*Sign* E.					

In this table it will be seen that there are five different kinds of units used in the currency of the United States, viz: *Mills, Cents, Dimes, Dollars,* and *Eagles;* and that they increase in value in a *tenfold* proportion. Accounts, however, are kept only in *dollars* and *cents.*

E

Q. Repeat the United States currency table. How many different units are there in it? Write down the signs used to represent them. How many mills in 1 cent? How many cents in 1 dime? How many mills in 1 dime? How many dimes in 1 dollar? How many cents in 1 dollar? How many mills? How many dollars in 1 eagle? How many dimes? Cents? Mills? In which of these units are accounts kept?

102. ENGLISH CURRENCY, OR STERLING MONEY.

					£	sh.	d.	far.
4 Farthings	make	1 Penny.	*Sign* D.		1	=20	=240	=960
12 Pence	"	1 Shilling.	*Sign* Sh.			1	=12	=48
20 Shillings	"	1 Pound.	*Sign* £				1	=4
21 Shillings	"	1 Guinea.						

This table is used in England, and was in common use in this country before the Revolution. It has 5 different units, viz: *Farthings, Pence, Shillings, Pounds,* and *Guineas.* Farthings are usually written in fractional parts of pence, thus: 1 far.=¼d.; 3 far.=¾d.

Q. Repeat the English currency table. What else is it called? What are the different kinds of units used? Write their signs? Where is this table used? Was it ever used in this country? How many farthings in 1 penny? How many pence in 1 shilling? How many farthings in 1 shilling? How many shillings in 1 pound? How many pence in 1 pound? How many farthings? How are farthings usually expressed? 3 farthings are equal to what?

103. AVOIRDUPOIS WEIGHT.

16 drachms	make 1 ounce.	*Sign* oz
16 ounces	" 1 pound.	*Sign* lb.
28 pounds	" 1 quarter.	*Sign* qr.
4 quarters or 112 pounds	" 1 hundred weight.	*Sign* cwt.
20 hundred weight	" 1 ton.	*Sign* T.

T.	cwt.	qr.	lb.	oz.	drachms.
1	=20	=80	=2240	=35840	=573440
	1	=4	=112	=1792	=28672
		1	=28	=448	=7168
			1	=16	=256
				1	=16

By this table heavy and coarse articles are weighed, such as meat, hay, groceries, and all the metals except gold and silver. It has *six* different *units,* viz: *tons, hundred weight, quarters, pounds, ounces* and *drachms.* Although by this *table 112 pounds* constitute the hundred weight, and 2240 lbs. the ton, in the State of New York and some of the other States the lawful hundred weight is 100 lbs. and the ton 2000 pounds.

Q. Repeat avoirdupois weight. What articles are weighed by it? How many kinds of units in it? What are they? How many pounds in a quarter? In a hundred weight? In a ton? Write the signs on your slate. Do the weights in New York correspond with this table? How many pounds make a ton in New York?

104. Troy Weight.

24 grains (gr.) make 1 pennyweight.   *Sign* pwt.
20 pennyweights "   1 ounce.   *Sign* oz.
12 ounces   "   1 pound.   *Sign* lb.

lb.	oz.	pwt.	gr.
1	=12	=240	=5760
	1	= 20	= 480
		1	= 24

By this table, gold, silver, and the other precious metals are weighed. There are *four* different units in it, viz: *pound, ounce, pennyweight* and *grain.*

Q. Repeat Troy weight. What articles are weighed by it? How many kinds of units in it? What are they? How many grains in a pennyweight? How many in an ounce? How many in a pound? How many pennyweights in an ounce? How many in a pound?

105. Apothecaries' Weight.

20 grains make 1 scruple.   *Sign* Ɵ
3 scruples "   1 drachm.   *Sign* 3
8 drachms "   1 ounce.   *Sign* 3
12 ounces "   1 pound.   *Sign* ℔

℔	3	3	Ɵ	gr.
1	=12	=96	=288	=5760
	1	= 8	= 24	= 480
		1	= 3	= 60
			1	= 20

This weight is used by apothecaries in compounding their medicines. It only differs from Troy weight in the different divisions of the ounce. The pound and ounce are the same *in the two tables.* When medicines are sold in large quan-

tities, the avoirdupois weight is used.   There are 5 different
units in apothecaries' weights, viz : *pound, ounce, drachm,
scruple* and *grain*.

Q. Repeat apothecaries' weight. How used ? Does it differ much
from Troy weight ? What units are the same in both tables ? Are
medicines always weighed by this table ? What other table is used ?
When ? How many kinds of units in this table ? What are they ?
How many ounces in a pound ? How many drachms ? How man
scruples ? How many grains ? Write the signs of the units.

106.   CLOTH MEASURE.

2¼ inches make 1 nail.   *Sign* n.
4 nails   "   1 quarter.   *Sign* q.
4 quarters "   1 yard.   *Sign* yd.

$$yd. \quad q. \quad n. \quad in.$$
$$1 = 4 = 16 = 36$$
$$1 = 4 = 9$$
$$1 = 2\tfrac{1}{4}$$

By this measure dry goods are sold in the United States.
The English, French, and Flemish measures differ from
these, and are used in the measurement of dry goods imported from these countries. There are *three quarters in
the Flemish ell,* 5 *quarters in the English ell,* and 6 *quarters in the French ell.*   In the United States cloth measure
there are 4 units : *yard, quarter, nail* and *inch.*

Q. Repeat cloth measure. For what purpose used ? Do we import
by this measure ? How many quarters in the Flemish ell ? In the
English ell ? In the French ? How many kinds of unit in United States
cloth measure ? What are they ? How many quarters in a yard ? How
many nails in a yard ? How many inches ? Write the signs.

107.   LONG MEASURE.

3 barleycorns	make 1 inch.	*Sign* in.
12 inches	" 1 foot.	*Sign* ft.
3 feet	" 1 yard.	*Sign* yd.
5½ yards	" 1 rod, perch or pole.	*Sign* r. or p.
40 rods	" 1 furlong.	*Sign* f.
8 furlongs	" 1 mile.	*Sign* m.
3 miles	" 1 league.	*Sign* L.
60 geographical or } 69½ statute miles } degrees	" 1 degree.	*Sign* deg. or °
	" The circumference of the earth.	

m.	l.	r. or p.	yd.	ft.	in.
1=8=	320=	1760	=5280	=63360	
	1=	40=	220	= 660	= 7920
		1=	5½=	16½=	198
			1 =	3 =	36
				1 =	12

By this table distances are measured. The units are *degree, league, mile, furlong, rod, yard, foot, inch* and *barleycorn.*

Q. Repeat long measure. For what purpose used? What are the units in it? How many furlongs in a mile? How many yards? How many feet? How many feet in a yard? How many inches? How many yards in a rod or pole? How many feet? Write down the signs.

108.   LAND OR SQUARE MEASURE.

144 square inches make 1 square foot.	*Sign* sq. ft.
9 square feet " 1 square yard.	*Sign* sq. yd.
30¼ square yards " 1 square pole.	*Sign* P.
40 square poles " 1 rood.	*Sign* R.
4 roods " 1 acre.	*Sign* A.
640 acres " 1 square mile.	*Sign* M.

A.	R.	P.	sq. yd.	sq. ft.	sq. in.
1=	4=	160=	4840	=43560	=6272640
	1=	40=	1210	=10890	=1568160
		1=	30¼=	272¼=	39204
			1 =	9 =	1296
				1 =	144

This table is used in measuring the content of ground or of any plane surface. A square foot is a square the sides of which are 1 foot in length; a square yard is a square the sides of which are 1 yard or 3 feet in length, &c. In measuring land, surveyors generally used an iron chain 66 feet in length, composed of 100 links, each link being 7.92 inches in length; 10 square chains make 1 acre. But their measures are reduced to those used in the above table. This table consists of the following units : *square mile, acre, rood, square pole, square yard, square foot, square inch.*

Q. Repeat land measure. For what purpose used? What is a square foot? A square yard? Square rod? Square mile? What measure do surveyors use in measuring land? How many feet in the surveyor's chain? How many links in it? What is the length of each link? Is the content of ground generally estimated in chains or in the land measure? What are the units used in land measure? How many acres in a square mile? How many roods in an acre? How many square poles in an acre? How many square yards? Write the signs. How many square chains make 1 acre?

109.   SOLID OR CUBIC MEASURE.

1728 cubic inches	make 1 cubic foot.
27 cubic feet	" 1 cubic yard.
40 cubic feet of round or 50 cubic feet of hewn timber	" 1 ton.
128 cubic feet	" 1 cord of wood.
42 cubic feet	" 1 ton of shipping.

By this table the solid contents of bodies are determined, as *stone, timber, earth, boxes of goods,* &c. A *cubic inch* is a solid with square faces, the sides of which are 1 inch in length. A *cubic foot* is a solid with square faces, the sides of which are 1 foot in length, &c. The *solid content of a body is ascertained by measuring its length, breadth, and thickness, and multiplying these measures together.* Thus, a *cubic foot* being a solid which is 12 inches long, 12 inches broad, and 12 inches thick, its solid content is $12 \times 12 \times 12 = 1728$ cubic inches, as in the table. And a *cubic yard* being a solid which measures 3 feet in each direction, its solid content will be $3 \times 3 \times 3 = 27$ cubic feet. If a box of dry goods measured 4 feet long, 2 feet broad, and 3 feet thick, its solid content would be found by multiplying these measures together—thus, $4 \times 2 \times 3 = 24$ *cubic feet.* A *cord of wood* measures 4 feet in depth, 4 feet in height, and 8 feet in width, and its solid content is $4 \times 4 \times 8 = 128$ cubic feet. The units in cubic measure are, *ton, cord, cubic yard, cubic foot,* and *cubic inch.*

Q. Repeat cubic measure. For what purpose used? What is a cubic inch? Cubic foot? Cubic yard? How is the solid content of a body ascertained? How many cubic inches does a cubic foot contain? Why? How many cubic feet does a cubic yard contain? Why? How would you ascertain the solid content of a box of goods? What would be the solid content of a box whose measures were 5 feet, 3 feet, and 8 feet? 8 yards, 9 yards, and 10 yards? What are the measures of a cord of wood? How many cubic feet in a cord of wood? Why? What are the units in cubic measure?

110.   LIQUID MEASURE.

4	gills	make 1 pint.	*Sign* pt.
2	pints	" 1 quart.	*Sign* qt.
4	quarts	" 1 gallon.	*Sign* gl.
31½	gallons	" 1 barrel.	*Sign* bbl.
2	barrels	" 1 hogshead.	*Sign* hhd.
2	hogsheads	" 1 pipe.	*Sign* p.
2	pipes	" 1 tun.	*Sign* T.

T.	p.	hhd.	bbl.	gl.	qt.	pt.	gills.
1=	2=	4=	8=	252	=1008	=2016	=8064
	1=	2=	4=	126	= 504	=1008	=4032
		1=	2=	63	= 252	= 504	=2016
			1=	31½=	126=	252	=1008
				1 =	4=	8=	32
					1=	2=	8
						1=	4

This measure is used for measuring liquids. The gallon contains 231 cubic inches. The units in this table are, *tun, pipe, hogshead, barrel, gallon, quart, pint* and *gill*.

Q. Repeat liquid measure. How many cubic inches does the gallon contain? What are the units in this table? How many gallons in a barrel? In a hogshead? In a pipe? In a tun? Write the signs.

111.   DRY MEASURE.

2 pints	make 1 quart.	*Sign* qt.
4 quarts	" 1 gallon.	*Sign* gl.
2 gallons	" 1 peck.	*Sign* pk.
4 pecks	" 1 bushel.	*Sign* bu.

bu.	pk.	gl.	qt.	pt.
1 =	4 =	8 =	32 =	64
	1 =	2 =	8 =	16
		1 =	4 =	8
			1 =	2

By this table grain, fruits, roots, salt and coal, &c., are measured. Coal is sold in some of the States by the ton weight, and also by the chaldron, which contains 36 bushels. The bushel in common use in the United States is the English *bushel, and measures* 18½ inches in diameter, and is inches deep. A bushel of this size contains 2150.42 cubic inches; and the gallon, which is ⅛ of the bushel, contains 268⅘ cubic inches. A barrel of flour contains 196 lbs.; the barrel for measuring unshelled corn 5 bushels. The units in this table are, *bushel, peck, gallon, quart* and *pint*.

Q. Repeat dry measure. For what purposes used? How is coal measured? What bushel is used in the United States? What are its dimensions? What are the cubic contents of the dry bushel? Of the dry gallon? What units in this table? What signs? How much does a barrel of flour weigh? How many bushels in a corn barrel?

112.   TIME.

60 seconds	make 1 minute.	*Sign* m.
60 minutes	" 1 hour.	*Sign* h.
24 hours	" 1 day.	*Sign* d.
7 days	" 1 week.	*Sign* w.
4 weeks	" 1 month.	*Sign* mo.
12 calendar, or 13 lunar or tabular months, 1 day, & 6 hours, or 365 days	" 1 common year	*Sign* yr.

yr.	w.	d.	h.	m.	s.
1=	52=	365¼=	8766=	525960=	31557600
	1=	7 =	168=	10080=	604800
		1 =	24=	1440=	86400
			1=	60=	3600
				1=	60

The calendar months are, 1. January, which has 31 days; 2. February, 28 days; 3. March, 31 days; 4. April, 30 days; 5. May, 31 days; 6. June, 30 days; 7. July, 31 days; 8. August, 31 days; 9. September, 30 days; 10. October, 31 days; 11. November, 30 days; 12. December, 31 days; in all, 365 days. As the common year contains 365¼ days, 1 day is added to February every 4th year, which makes up the deficiency in the calendar year. The units are, *year, month, week, day, hour, minute* and *second*.

Q. Repeat the table for time. What are the calendar months? How many days in each? Do the calendar and common year correspond? What difference? How is the difference made up? What are the units in this table? How many days in a week? How many days in a common year? What units in this table?

113. CIRCULAR MEASURE.

60 seconds *sign* ″ make 1 minute.   *Sign* ′.
60 minutes   " 1 degree.   *Sign* °.
30 degrees   " 1 sign.   *Sign* S.
12 signs or 360 degrees " 1 circumference of a circle. *Sign* cir.

$$\begin{array}{ccccc}
\text{Cir.} & \text{S.} & ° & ' & '' \\
1 = & 12 = & 360 = & 21600 = & 1296000 \\
& 1 = & 30 = & 1800 = & 108000 \\
& & 1 = & 60 = & 3600 \\
& & & 1 = & 60
\end{array}$$

By this measure the arcs of circles are measured. It is used in *Astronomy* and *Geography*. The distances and motions of the heavenly bodies in astronomy being measured by arcs of circles, and the distances between places on the surface of the earth by their latitude and longitude, which are estimated on the arcs of circles. The units in this table are *circumference, sign, degree, minute* and *second*.

Q. Repeat circular measure. For what purposes used? How is it used in astronomy? In geography? What units in this table? How many degrees in a circumference?

114. MISCELLANEOUS TABLES.

(1) 12 things make 1 dozen.   *Sign* doz.
    12 dozen   "   1 gross.   *Sign* gro.
    12 gross   "   1 great gross.   *Sign* g. gro.

$$\begin{array}{cccc}
\text{G. gro.} & \text{gro.} & \text{doz.} & \text{things.} \\
1 = & 12 = & 144 = & 1728 \\
& 1 = & 12 = & 144 \\
& & 1 = & 12
\end{array}$$

    20 things   make 1 score.
    24 sheets of paper   "   1 quire.
    20 quires   "   1 ream.

(3) A folio book has each
    sheet folded in   2 leaves, which make 4 pages.
    A quarto or 4to book   4 "   "   " 8 "
    An octavo or 8vo   8 "   "   " 16 "
    A duodecimo or 12mo 12   "   "   " 24 "
    An octodecimo or 18mo 18   "   "   " 36 "

E 2

Q. How many things make a dozen? How many dozen a gross? How many gross a great gross? How many things in a gross? How many things in a score? How many sheets of paper in a quire? How many quires in a ream? How is a folio book folded? What is a quarto book? An 8vo? A 12mo? An 18mo? How many pages on a sheet when in folio? When in 4to? When in 8vo? When in 12mo? When in 18mo?

OF REDUCTION OF COMPOUND NUMBERS.

115. To prepare Compound Numbers for the various operations of arithmetic, it is often necessary that they should undergo certain reductions.

REDUCTION *of Compound Numbers consists in changing the kind of unit in which the numbers are expressed without altering their value.*

Thus, 20 dollars and 15 cents are expressed in terms of different units, viz. *dollars* and *cents ;* but since 1 dollar contains 100 cents (Art. 101), 20 dollars contain 20×100 =2000 cents. Hence, 20 dollars and 15 cents will be equal to 2000 cents + 15 cents = 2015 cents. Here the given numbers have been expressed in terms of the same unit without their value being altered, and this has been done by reducing the dollars to their equivalent value in terms of a *lower* unit, viz. *cents.*

Again, since 20 shillings make 1 pound (Art. 102), 30 shillings are equal to $\frac{30}{20}$ pounds, or 1 pound and 10 shillings. Here the shillings have been expressed in terms of a *higher* unit and shillings.

Hence, Reduction of Compound Numbers may be effected in two ways,

I. *By changing their value from a higher to a lower unit, called Reduction Descending.*

II. *By changing their value from a lower to a higher unit, called Reduction Ascending.*

Q. For what purposes are reductions necessary in compound numbers? In what does reduction of compound numbers consist? In how many ways may the reduction be effected? What are they? What is reduction ascending? Descending? When 20 dollars and 15 cents are expressed in cents, will it be reduction ascending or descending? How many cents in 20 dollars and 15 cents? When 30 shillings are expressed *in pounds* and shillings, will it be reduction ascending or descending? *How many pounds* and shillings in 30 shillings?

OF REDUCTION DESCENDING.

116. In Reduction Descending the value of the numbers is changed from a higher to a lower unit.

Thus, reduce 10 pounds 5 shillings to shillings.

Since 20 shillings make 1 pound (Art. 102), by multiplying 10 pounds by 20, we bring all the pounds to shillings; and then adding in the 5 shillings, we reduce the 10 pounds and 5 shillings to their equivalent value, 205 shillings.

```
OPERATION.
10  pounds.
20
―――
200 shillings.
  5 shillings.
―――
205 shillings.
```

If it had been required to bring the given number to *pence*, since 12 pence make 1 shilling, (art. 102), the number of pence in 205 shillings will be equal to 205 × 12 = 2460 pence.

Hence, for reduction descending we have the following

RULE.

Multiply the number expressed in terms of the highest unit, by as many of the next lower unit as make 1 of the higher; and to this product add the number belonging to the lower unit. Proceed in this way through all the different units to the lowest,—the last sum will be the required number.

Q. What is reduction descending? What is the rule?

117. UNITED STATES CURRENCY OR FEDERAL MONEY.

Reduce $110, 99 cts. and 9 mills to mills.

In this example we multiply the number expressed in terms of the highest unit, which is 110 *dollars*, by 100, the number of cents in 1 dollar, and to this product add the number of cents, 99. This gives us 11099 cents. Multiplying again by 10, the number of mills in 1 cent, and to this product adding in the 9 mills, the given number contains **110999** *mills.*

```
OPERATION.
110   dollars.
100
―――
11000  cents.
   99  cents.
―――
11099  cents.
    10
―――
110990 mills.
     9 mills.
―――
110999 mills.
```

Note.—Since the units in the United States currency increase in value in a tenfold ratio, it is evident that eagles may be converted into dollars by annexing 1 cipher; into dimes by annexing 2; into cents 3; and into mills 4 ciphers; and that dollars are changed into cents by annexing 2 ciphers, and into mills 3.

Thus, 5 eagles=$50=500 d.=5000 cts.=50000 mills.

      5 dollars=500 cents=5000 mills.

EXAMPLES.

1. Reduce 15 eagles, $5 and 10 cts. to mills.

                   Ans. 155100 mills.

2. Reduce $325, 25 cents, 4 mills, to mills.

                   Ans. 325254 mills.

3. Reduce $89, 4 mills, to mills.   Ans. 89004 mills.

4. Reduce $750, 99 cents, to cents.   Ans. 75099 cents.

5. Reduce 1000 eagles to cents, and then to mills.

6. How many cents in $1000 and 10 cents?

Q. Repeat the United States currency table. How do the units increase in value in this table? How then may eagles be brought to dollars? To dimes? To cents? To mills? How may dollars be brought to cents? To mills?

118. ENGLISH CURRENCY OR STERLING MONEY.

Reduce £5 17s. 6d. to pence.

Here we multiply the £5 by 20 to bring them to shillings, and add in mentally the 17 shillings. Multiplying now by 12, and adding in mentally the 6 pence, 1410 pence is the result.

OPERATION.

$$
\begin{array}{r}
£5 \quad 17s. \quad 6d. \\
20 \\
\hline
117 \text{ shillings.} \\
12 \\
\hline
1410 \text{ pence.}
\end{array}
$$

EXAMPLES.

1. Reduce £100 to pence.   Ans. 24000 pence.

2. Reduce £1234 15s. 7d. to farthings.

                   Ans. 1185388 farthings.

3. Reduce £1 0s. 7½d. to farthings.

                   Ans. 990 farthings.

4. Reduce £1 10s. to shillings.   Ans. 30 shillings.

5. Reduce 100 guineas to farthings.

                   Ans. 100800 farthings.

6. Reduce 12s. 6d. to pence.   Ans. 150 pence.

Q. Repeat English currency table (Art. 102).

119. AVOIRDUPOIS WEIGHT.

Reduce 15 tons, 8 cwt. 2 qrs. 12 lbs. 8 oz. 7 drs. to drachms.

OPERATION.

In this example we multiply the 15 tons by 20 to bring them to cwt. and add in the 8 (Art. 103), then by 4, and add in the 2 qr., then by 28, and add in the 12 lb., then by 16, and add in the 8 oz., then by 16, and add in the 7 drachms. The result will express the number of drachms in the given numbers.

T.	cwt.	qr.	lb.	oz.	dr.
15	8	2	12	8	7

20

308 cwt.
4

1234 qrs.
28

34564 lbs.
16

553032 oz.
16

8848519 dr.

EXAMPLES.

1. Reduce 35 tons, 17 cwt. 1 qr. 23 lb. 7 oz. 18 dr. to drs.
Ans. 20571005 drachms.

2. Reduce 15 tons, 4 cwt. 14 lbs. to pounds.
Ans.

3. Reduce 20 lb. 6 oz. 3 dr. to drachms.
Ans.

4. Reduce 2 cwt. 4 oz. 3 dr. to drachms.
Ans.

5. Reduce 3 cwt. 12 lbs. 1 oz. to ounces.
Ans.

6. Reduce 16 lb. 11 oz. to drachms.
Ans.

7. Reduce 11 oz. 15 drs. to drachms.
Ans.

8. Reduce 15 oz. to drachms. Ans.

Q. Repeat avoirdupois weight (Art. 103). What do you multiply by to bring tons to cwt.? To qrs.? To lbs.? To oz.? To dr.?

120.   TROY WEIGHT.

Reduce 15 lbs. 3 oz. 2 pwts. 7 gr. to grains.

In this example we multiply the 15 lb. by 12 to bring them to oz. (Art. 104), and add in the 3 oz., then multiply by 20 to bring the whole to pwts., and add in the 2 pwts., then multiply by 24, and add in the 7 grains, and the result is the number of grains in the given number.

OPERATION.

lbs.	oz.	pwt.	gr.
15	3	2	7
12			

183 oz.
20

3662 pwt.
24

87895 grains.

EXAMPLES.

1. Reduce 25 lbs. 0 oz. 15 pwt. 16 gr. to grains.
                                        Ans. 144376 grains.
2. Reduce 16 lbs. 11 oz. to ounces.   Ans.
3. Reduce 10 lbs. to grains.   Ans.
4. Reduce 7 oz. 12 pwt. to pwt.   Ans.

Q. Repeat Troy weight (Art. 104). By what do you multiply to bring lbs. to oz.? To pwt.? To gr.?

121.   APOTHECARIES' WEIGHT.

Reduce 5 ℔, 7 ℥, 4 ʒ, 2 Ɖ, 11 gr. to grains.

In this example we multiply the 5 lb. by 12 to reduce them to ounces (Art. 105), and add in the 7 ounces; we then multiply by 8 to bring this result to drachms, and add in the 4 drachms; we then multiply by 3 to obtain scruples, and add in the 2 scruples; and finally by 20, and add in the 11 to produce rains.

OPERATION.

℔	℥	ʒ	Ɖ	gr.
5	7	4	2	11
12				

67 ounces.
8

540 drachms.
3

1622 scruples.
20

32451 grains.

1. Reduce 15 ℔, 0 ℥, 5 ʒ, to drachms.   Ans. 1445 drachms.
2. Reduce 6 ℔, 3 ℥, 2 ʒ, 1 ϶, to scruples.
                                                   Ans.
3. Reduce 8 ℔, 1 ℥, 6 ʒ, to drachms.   Ans.

Q. Repeat Apothecaries' weight (Art. 105). How do you reduce pounds to ounces ? To drachms ? To scruples ? To grains ?

122.   CLOTH MEASURE.

Reduce 12 yds. 2 qrs. 3 na. to nails.

We reduce the 12 yards to quarters by multiplying by 4 (Art. 106), and add in the 2 quarters; then to nails by multiplying by 4, and adding in the 3 nails.

OPERATION.

yds.	qrs.	na.
12	2	3
4		

50 quarters.
4

203 nails.

1. Reduce 25 yds. 3 qrs. to nails.   Ans. 412 nails.
2. Reduce 120 yds. to qrs.   Ans.
3. Reduce 15 yds. 0 qrs. 2 na. to inches.   Ans.

Q. Repeat cloth measure (Art. 106). How are yards reduced to quarters ? To nails ? How many nails in 5 yards ? How many inches in 1 yard ?

123.   LONG MEASURE.

Reduce 20 m. 3 fur. 20 R. 3 yds. to feet.

OPERATION.

m.	fur.	R.	yds.
20	3	20	3
8			

In this example we multiply the 20 miles by 8 to bring them to furlongs (Art. 107), and add in the 3 furlongs; then multiply by 40 to obtain rods, and add in the 20 rods; then by $5\frac{1}{2}$ to reduce the result to yards, and add in the 3 yards; and finally by 3 to obtain feet.

163 furlongs.
40

6540 rods.
$5\frac{1}{2}$

35973 yards.
3

107919 feet.

1. Reduce 365 miles to inches.   Ans. 23126400 inches.
2. Reduce 27 yds. 2 ft. to barleycorns.
                                    Ans.
3. Reduce 35 rds. 10 yds. 0 ft. 3 in. to inches.
                                    Ans.

Q. Repeat long measure (Art. 107). How are miles reduced to feet? How are rods reduced to inches? How are yards reduced to feet? To inches?

124.   LAND OR SQUARE MEASURE.

Reduce 12 M. 20 A. 2 R. to square yards.

OPERATION.

M.   A.   R.
12   20   2
640
———
7700 acres.
4
———
30802 roods.
40
———
1232080 square poles.
$30\frac{1}{4}$
———
37270420 square yards.

To bring the 12 miles to acres we multiply by 640 (Art. 108), and add in the 20 acres; then multiply by 4; then by 40 and by $30\frac{1}{4}$ to bring the whole to square yards.

1. Reduce 24 A. 3 R. 35 P. to square yards.
                                Ans. 120848 yds.
2. Reduce 100 M. 7 A. 0 R. 20 P. to square rods.
                                Ans.
3. Reduce 177 acres to square inches.   Ans.

Q. Repeat land measure. How are miles reduced to square inches? How are acres reduced to square inches?

125. SOLID OR CUBIC MEASURE.

Reduce 14 tons of round timber to cubic inches.

OPERATION.

Since 40 cubic feet of round timber make 1 ton (Art. 109), we multiply the 14 by 40, and afterwards
multiplying this product by 1728, we
reduce the whole to cubic inches.

```
T.
 14
 40
----
560 cubic feet.
1728
----
967680 cubic inches.
```

EXAMPLES.

1. Reduce 27 cords of wood to cubic feet.   Ans. 3456
2. How many cubic inches in 47 cords 5 cubic feet of wood?
                                                Ans.
3. How many cubic inches in 200 tons shipping?
                                                Ans.
4. How many cubic inches in 40 cubic yards and 12 cubic
   feet?   Ans.

Q. Repeat cubic measure (Art. 109). How are tons reduced to cubic
inches? How are shipping tons reduced to cubic inches? How do you
find the number of cubic feet in a cord of wood?

126. LIQUID MEASURE.

Reduce 4 tuns to gallons.

OPERATION.

By Art. 110, 2 pipes make 1
tun; we then multiply the 4 by 2;
then this result by 2 again to bring
it to hogsheads; then by 2 again
to bring hogsheads to barrels; and
finally by 31½ to reduce this result
to gallons.

```
T.
 4
 2
----
 8 pipes.
 2
----
16 hogsheads.
 2
----
32 barrels.
31½
----
Ans. 1008 gallons.
```

EXAMPLES.

1. Reduce 12 pipes, 1 hhd. 0 bbls. 14 galls. to pints.

Ans. 12712.

2. Reduce 3 hhds. 4 galls. to quarts.   Ans.

3. Reduce 17 bbls. 4 quarts to gills.   Ans.

Q. Repeat liquid measure (Art. 110). How would you find the number of gallons in a tun? The number of pints in a hogshead? In a barrel?

127.   DRY MEASURE.

Reduce 16 bushels, 2 pecks, 1 gallon, to quarts.

OPERATION.

In this example we multiply the 16 bushels by 4 to bring them to pecks (Art. 111), and add in the 2 pecks; then multiply by 2 and by 4 to reduce the whole to quarts.

bush. peck gall.

16   2   1

4

66 pecks.

2

133 gallons.

4

532 quarts.

EXAMPLES.

1. Reduce 17 bushels, 2 pecks, and 3 quarts, to pints.

Ans. 1126

2. Reduce 40 bush. 1 gal. to quarts.   Ans.

3. Reduce 27 chaldrons and 4 bushels of coals to pecks.

Ans.

Q. Repeat dry measure (Art. 111). How are bushels reduced to pecks? To gallons? To quarts? To pints? How is a chaldron of coal reduced to pecks?

128. TIME.

Reduce 250 days, 5 hours, 13 minutes, to seconds.

OPERATION.

days. hrs. m.
250 5 13
24

In this example we multiply the 250 days by 24 to bring them to hours (Art. 112); this result by 60 to bring it to minutes; and then by 60 to bring the whole to seconds.

6005 hours.
60

360313 minutes.
60

21618780 seconds.

EXAMPLES.

1. Reduce 14 years to minutes.   Ans. 7358400
2. Reduce 50 weeks 6 days to hours.   Ans.
3. Reduce 11 months, 3 weeks, 5 days, 3 hours, to seconds.   Ans.

Q. Repeat the table for time (Art. 112). How are months reduced to hours? To seconds?

129. CIRCULAR MEASURE.

Reduce 40° 3′ 27″ to seconds.

OPERATION.

40° 3′ 27″
60

Here we multiply by 60 to obtain minutes, and then by 60 to reduce to seconds (Art. 113).

2403
60

144207 seconds.

EXAMPLES.

1. Reduce 360° to seconds.   Ans. 1296000
2. Reduce 14° 0′ 34″ to seconds.   Ans.
3. Reduce 118° 1′ to minutes.   Ans.

Q. Repeat circular measure (Art. 113). How are degrees reduced to seconds?

REDUCTION ASCENDING.

130. *In Reduction Ascending the value of the numbers is changed from a lower to a higher unit.*

Thus, reduce 4135 farthings to pounds sterling.

Since 4 farthings make 1 penny, if we divide 4135 farthings by 4, the quotient will be the number of pence which they contain. They contain 1033 pence and 3 farthings over. Dividing now the pence by 12, we shall have the number of shillings in 1033 pence, since 12 pence make 1 shilling. There are 86 shillings

OPERATION.

4) 4135 farthings.

12) 1033 — 3 far. left.

2,0) 8,6 — 1 penny left.

   4 — 6 sh. left.

Ans. £4 6s. 1¾d.

and 1 penny over. Dividing now the shillings by 20, we find the number of pounds in 86 shillings, since 20 shillings make 1 pound. There are 4 pounds and 6 shillings over. There are therefore £4 6s. 1¾d. in 4135 farthings.

As the same reasoning will equally apply to numbers expressed in terms of any kind of unit, we have for reduction ascending the following

RULE.

Divide the given number by as many of its units as make 1 of the next higher unit. Continue to divide each successive quotient by the number of units of the lower in the next higher unit, the last quotient will be the answer. Should there be remainders, they must be set aside and connected with the last quotient.

Q. What is reduction ascending? How are farthings brought to pence? How are pence brought to shillings? How are shillings brought to pounds? What is the rule for reduction ascending?

131. UNITED STATES CURRENCY OR FEDERAL MONEY.

Reduce 27543 mills to dollars.

Since 10 mills make 1 cent, by dividing 27543 by 10, the quotient will be cents : there are 2754 cents and 3 mills over. Dividing this quotient by 100, since 100 cents make 1 dollar, we find the number of dollars. There are 27 dollars and 54 cents over. The answer is $27, 54 cts. 3 m.

OPERATION.

1,0) 2754,3 mills.

1,00) 27,54 — 3 mills over.

27 — 54 cents over.

Ans. $27 54 cts. 3 m.

Note. Since in United States currency the divisors are 10, 100, &c., the quotient may be at once obtained as in like divisions of simple numbers (Art. 54), by cutting off from the right one figure to reduce mills to cents, two figures to reduce cents to dollars, &c.

EXAMPLES.

1. Reduce 17547 mills to dimes.   Ans. 175 d. 4 cts. 7 m.
2. Reduce 2754 cents to eagles.   Ans. E 2 $7 54 cts.
3. Reduce 145430 cents to dollars.   Ans. $1454 30 cts.
4. Reduce 100400 mills to dollars.   Ans. $100 40 cts.

Q. Repeat United States currency table (Art. 101). How may mills be at once reduced to cents ? Cents to dollars ? How many dollars in 100 cents ? How many eagles in 100 dollars ?

132. ENGLISH CURRENCY OR STERLING MONEY.

Reduce 13476 pence to pounds sterling.

Dividing the pence by 12 to bring them to shillings, and the shillings by 20 to bring them to pounds, the result is £56 3s. 0d.

OPERATION.

12) 13476

2,0) 112,3

56. 3   Ans. £56 3s. 0d.

EXAMPLES.

1. Reduce 17463 farthings to pounds.   Ans. £18 3s. 9¾d.
2. Reduce 3046 shillings to pounds.   Ans.
8. Reduce 1764 farthings to shillings.   Ans.

Q. Repeat English currency table. How are farthings brought to pence ? Pence to shillings ? Shillings to pounds ?

133.   AVOIRDUPOIS WEIGHT.

Reduce 7654304 drachms to tons.

We first divide the drachms by 16 to reduce them to ounces, since 16 drachms make 1 ounce (Art. 103); then this result by 16 again to bring the ounces to pounds; then the pounds by 28 to bring them to quarters; the quarters by 4 to bring them to hundred weight; and the hundred weight by 20 to get tons.

OPERATION.

16) 7654304 drachms.

16) 478394

28) 29899 — 10 ounces left.

4) 1067 — 23 pounds left.

2,0) 26,6 — 3 quarters left.

13. — 6 cwt. left.

Ans. 13 T. 6 cwt. 3 qr. 23 lb. 10 oz.

Since 16=4×4 and 28=7×4, the divisions might have been simplified by dividing by the factors which compose these numbers.

EXAMPLES.

1. Reduce 103460 ounces to hundred weight.   Ans. 57 cwt. 2 qr. 26 lb. 1 oz.
2. Reduce 5043720 drachms to tons.   Ans.
8. Reduce 14767 pounds to tons.   Ans.

Q. Repeat avoirdupois weight. How are drachms brought to ounces ? *Ounces to pounds ? Pounds to quarters ? Quarters to hundred weight ? Hundred weight to tons ?* How may the divisions be simplified when *the* divisors are 16 and 28 ?

134. TROY WEIGHT.

Reduce 6049 grains to pounds.

The grains are brought to pennyweights by dividing by 24 or by 6 and 4, since $6 \times 4 = 24$; the pennyweights are reduced to ounces by dividing by 20, and the ounces to pounds by dividing by 12 (Art. 104).

OPERATION.

24) 6049 grains.

2,0) 25,2 — 1 grain left.

12) 12 — 12 pwt. left.

1

Ans. 1 lb. 0 oz. 12 pwt. 1 gr.

EXAMPLES.

1. Reduce 23465 grains to pounds.

Ans. 4 lb. 0 oz. 17 pwt. 17 gr.

2. Reduce 10376 pennyweights to pounds.

Ans.

3. Reduce 10364 grains to pounds.

Ans.

Q. Repeat Troy weight. How are grains brought to pennyweights? Pennyweights to ounces? Ounces to pounds?

135. APOTHECARIES' WEIGHT.

Reduce 34769 grains to pounds.

OPERATION.

2,0) 3476,9 grains.

3) 1738—9 gr. left.

8) 579—1 scruple left.

12) 72—3 drachms left.

6

We divide the grains by 20 to bring them to scruples; the scruples by 3 to bring them to drachms; the drachms by 8 to bring them to ounces; and the ounces by 12 to bring them to pounds (Art. 105).

Ans. 6℔ 0℥ 8ʒ 1Ə 9 gr.

1. Reduce 5046 drachms to pounds.   Ans. 52℔ 6℥ 6ℨ
2. Reduce 16134 scruples to pounds.   Ans.
3. Reduce 404376 grains to pounds.   Ans.

Q. Repeat apothecaries' weight. How are grains brought to scruples? Scruples to drachms? Drachms to ounces? Ounces to pounds?

136.   CLOTH MEASURE.

Reduce 1409 nails to yards.

We divide the nails by 4 to bring them to quarters, and the quarters by 4 to bring them to yards (Art. 106).

OPERATION.

4) 1409 nails.
————
4) 352 — 1 na. left.
————
88

Ans. 88 yd. 0 qr. 1 na.

1. Reduce 304 quarters to yards.   Ans. 76 yards.
2. Reduce 1093 nails to yards.   Ans.
3. How many ells English and Flemish in 390 nails?   Ans.

Q. Repeat cloth measure. How are nails brought to quarters? Quarters to yards? Quarters to ells English? To ells Flemish?

137.   LONG MEASURE.

Reduce 17643 barleycorns to yards.

We divide the barleycorns by 3 to bring them to inches; the inches by 12 to bring them to feet; and the feet by 3 to bring them to yards. Had it been required to bring the yards to rods, we should have to divide by 5½, since 5½ yards *make 1 rod.* To make this *ivision* we multiply the 163

OPERATION.

3) 17643 barleycorns.
————
12) 5881.
————
3) 490 — 1 in. left.
————
163 — 1 ft. left.

Ans. 163 yd. 1 ft. 1 in.

yards by 2 to bring them to *halves;* and at the same time reduce 5¼ to halves. There are 11 halves in 5¼. Dividing now the number of halves in 163 by 11, the quotient will be the rods required (Art. 85).

EXAMPLES.

1. Reduce 10476342 barleycorns to miles.

Ans. 55 m. 0 f. 36 p. 5 yds. 0 f. 6 in.

2. Reduce 90436 feet to miles.

Ans.

3. Reduce 2769 yds. to miles.

Ans.

4. Reduce 147694 yards to degrees.

Ans.

Q. Repeat long measure (Art. 107). How are barleycorns brought to inches? Inches to feet? Feet to yards? Yards to rods? Rods to furlongs? Furlongs to miles? Miles to leagues? Miles to degrees?

138. LAND OR SQUARE MEASURE.

Reduce 76033 square inches to square yards.

We divide the square inches by 144 to bring them to square feet, and the square feet by 9 to bring them to sq. yards (Art. 108.) Had it been required to bring the square yards to square poles, we should have to divide them by 30¼; to do which the yards are brought to quarters by multiplying by 4, and 30¼ is reduced to quarters also: their quotient will give the square poles required (Art. 85).

OPERATION.

144) 76033 sq. inches.

9) 528 — 1 sq. in. left.

58 — 6 sq. ft. left.

Ans. 58 sq. yd. 6 sq. ft. 1 sq. in.

F

1. Reduce 190467 square yards to acres.

Ans. 39 A. 1 R. 16 P. 13 yd.

2. Reduce 304760 square feet to acres.

Ans.

3. Reduce 91064067 square yards to square miles.

Ans.

Q. Repeat land measure. How are square inches brought to square feet? Square feet to square yards? Square yards to square poles? Square poles to roods? Roods to acres? Acres to square miles?

139.   SOLID OR CUBIC MEASURE.

Reduce 270645 cubic inches to tons of round timber.

In this example we divide the cubic inches by 1728 to bring them to cubic feet, and the feet by 40 to bring them to tons of round timber (Art. 109).

OPERATION.

1728) 270645 cubic inches.

4,0) 15,6–1077 cu. in. left.

3 — 36 cu. ft. left.

Ans. 3 tons, 36 ft. 1077 in.

1. Reduce 10643 cubic feet to cords.   Ans. 83 cords 19 ft.
2. Reduce 1196 cubic yards to cords.   Ans.
3. Reduce 1764356 cubic inches to tons of hewn timber.

Ans.

Q. Repeat solid or cubic measure. How are cubic inches reduced to cubic feet? Cubic feet to cubic yards? To tons of round timber? To tons of hewn timber? To cords of wood?

140.   LIQUID MEASURE.

Reduce 19604 quarts to hogsheads.

OPERATION.

4) 19604 quarts.

63) 4901

77 — 50 gls. left.

We divide the quarts by 4 to bring them to gallons, and the gallons by 63 to bring them to hogsheads (Art. 110).

Ans. 77 hhds. 50 gals.

1. Reduce 106464 gills to barrels.   Ans. 105 bbls. 19½ gal.
2. Reduce 160437 quarts to tuns.   Ans.
3. Reduce 98064 gallons to pipes.   Ans.

Q. Repeat liquid measure. How are gills brought to quarts? Quarts to gallons? Gallons to barrels? To hogsheads? Hogsheads to pipes? Pipes to tuns?

141.   DRY MEASURE.

Reduce 16224 pints to bushels.

OPERATION.

We divide the pints by 2 to bring them to quarts; the quarts by 4 to bring them to gallons; the gallons by 2 to bring them to pecks; and the pecks by 4 to bring them to bushels (Art. 111).

2) 16224 pints.
4) 8112
2) 2028
4) 1014
253—2 pks. left.

Ans. 253 bu. 2 pks.

1. Reduce 1077 gallons to bushels.

Ans. 134 bu. 2 pk. 1 gal.

2. Reduce 648 bushels to chaldrons of coal.

Ans.

3. Reduce 4336 quarts to bushels.

Ans.

Q. Repeat dry measure. How are pints reduced to quarts? Quarts to gallons? To pecks? Pecks to bushels? Bushels to coal chaldrons?

142.   TIME.

Reduce 106433 minutes to days.

We divide the minutes by 60 to reduce them to hours, and the hours by 24 to reduce them to days (Art. 112).

OPERATION.

6,0) 10643,3 minutes.
————————
24) 1773—53 minutes lef'
————————
73—21 hours left.

Ans. 73 dy. 21 h. 53 m.

EXAMPLES.

1. Reduce 376504302 seconds to years of 365 days each.
        Ans. 11 y. 342 dy. 16 h. 31 m. 42 s.
2. Reduce 1060405 minutes to years of 365 days each.
        Ans.
3. Reduce 909907 minutes to weeks.
        Ans.

Q. Repeat time table. How are seconds reduced to minutes? Minutes to hours? Hours to days? Days to weeks? Weeks to months? Months to years? Days to years?

143.   CIRCULAR MEASURE.

Reduce 1064709 seconds to degrees.

We divide the seconds by 60 to bring them to minutes, and the minutes by 60 to bring them to degrees (Art. 113).

OPERATION.

6,0) 106470,9 seconds.
————————
6,0) 1774,5 — 9 sec. left.
————————
295 — 45 minutes.

Ans. 295° 45' 9".

EXAMPLES.

1. Reduce 196053 seconds to degrees.   Ans. 54° 27' 33"
2. Reduce 4065 minutes to degrees.   Ans.
3. Reduce 1965097 seconds to degrees.   Ans.

Q. Repeat circular measure. How are seconds brought to minutes? Minutes to degrees?

ADDITION OF COMPOUND NUMBERS.

144. ADDITION of Compound Numbers consists in finding the sum of several numbers expressed in terms of different units.

Thus, let it be required to add $14.39 cts. 3 m., $10.94.6, $17.45.5, $20.00.7.

We place dollars under dollars, cents under cents, &c., and commence as in addition of simple numbers on the right, and say 7 and 5 are 12 and 6 are 18 and 3 are 21; but since the numbers in the column we are adding are mills, the sum is 21 mills.

OPERATION.

$	cts.	m.
14	39	3
10	94	6
17	45	5
20	00	7
Ans. 62	80	1

But 10 mills make 1 cent; there are therefore 2 cents and 1 mill in 21 mills. Setting down the 1 mill under the column of mills, we carry 2 cents into the column of cents, and say, 2 we carry and 0 are 2 and 5 are 7 and 4 are 11 and 9 are 20; set down 0 and carry 2; 2 we carry and 0 are 2 and 4 are 6 and 9 are 15 and 3 are 18; the sum of the cents is then 180 cents; but, since 100 cents make 1 dollar, there are 1 dollar and 80 cents in 180 cents. Setting down the 80 under the column of cents, we carry 1 dollar to the column of dollars, and adding up in the same way, we find the sum to be $62.80.1.

Again, let it be required to find the sum of £17 6s. 4d.; £27 14s. 8d.; £19 17s. 9d.

Placing pounds under pounds, shillings under shillings, &c., we commence on the right, and say 9 and 3 are 12 and 4 are 16, which is 16 pence; but since 12 pence make 1 shilling, there is 1 shilling in 16 pence and 4 pence over.

OPERATION.

£	s.	d.
17	6	4
27	14	8
19	17	9
Ans. 64	18	4

Set down the 4 under the column of pence, and carry 1 shilling *to the column* of shillings. Adding up the shillings, we find their sum to be 38 shillings; but since 20 shillings make 1 pound, 38 shillings are equal to 1 pound 18 shillings; set down the 18 under the column of shillings, and carry 1 pound to the column of pounds. We find the sum to be £64 18s. 4d.

From these examples we deduce the following

RULE. [1]

I. *Set down units of the same kind under each other.*

II. *Beginning at the column on the right, find the sum of the units in this column, and divide this sum by as many of its units as make 1 of the next higher unit; set down the remainder, if there be any, under the column added up, and carry the quotient to the next column.*

III. *Proceed in this way with all the columns, and set down under the last column the entire sum.*

Addition of compound numbers is proved as in addition of simple numbers.

Q. What is addition of compound numbers? How are the numbers set down? Where do you begin to add? Do you set down the entire sum under the column added up? What do you set down? What do you carry? What is the rule? What is the proof for addition of compound numbers?

EXAMPLES.

145. UNITED STATES CURRENCY, OR FEDERAL MONEY.

(1)			(2)			(3)	
$	cts.	m	$	cts.	m,	$	cts.
104	07	3	1004	00	2	294	06¼
97	84	9	4076	94	0	180	18¾
146	30	2	3976	65	7	344	27
87	45	3	2009	05	9	108	87½
435	67	7	11066	65	8	927	39¼

4. Add $4.37½, $19.31¼, $100.33⅓.

5. Add $99.75, $104.62½, $75.18¾.

6. Add $105.06¼, $94.68¾, $109.94¼.

Q. Repeat United States currency table.

146. ENGLISH CURRENCY, OR STERLING MONEY.

(1)			(2)			(3)		
£	s.	d.	£	s.	d.	£	s.	d.
104	16	4	90	19	4⅓	104	13	1
37	04	8	36	05	3¼	94	19	11½
17	19	8	17	18	1¼	96	10	10¾
94	03	2	24	09	5½	87	6	4¾
254	3	5						

Q. Repeat English currency.

147. AVOIRDUPOIS WEIGHT.

(1)					(2)					(3)				
cwt.	(4) qr.	(28) lb.	(16) oz.	(16) dr.	cwt.	(4) qr.	(28) lb.	(16) oz.	(16) dr.	cwt.	(4) qr.	(28) lb.	(16) oz.	(16) dr.
17	2	24	15	10	3	3	27	15	4	99	2	3	14	1
29	1	17	13	9	27	1	1	0	6	194	2	14	11	9
34	3	27	10	15	94	3	24	11	9	67	0	13	0	0
104	3	12	9	14	64	2	19	10	10	14	9	06	3	9
186	3	27	2	0										

Q. Repeat avoirdupois weight.

148. TROY WEIGHT.

(1)				(2)				(3)			
lb.	(12) oz.	(20) dwt.	(24) gr.	lb.	(12) oz.	(20) dwt.	(24) gr.	lb.	(12) oz.	(20) dwt.	(24) gr.
94	11	17	21	191	0	19	23	471	10	14	3
100	9	13	10	46	3	2	19	970	9	19	23
43	11	16	23	30	11	18	17	94	3	10	10
19	01	14	20	91	9	08	16	105	11	4	22
258	11	3	2								

Q. Repeat Troy weight.

149.   APOTHECARIES' WEIGHT.

(1)

(12) ℔	(8) ℥	(3) ʒ	(20) Ə	gr.
10	9	3	2	16
9	2	7	1	19
13	11	6	2	14
34	0	2	1	9

(2)

(12) ℔	(8) ℥	(3) ʒ	(20) Ə	gr.
16	3	7	2	14
29	11	4	2	19
22	10	5	1	19

(3)

(12) ℔	(6) ℥	(3) ʒ	(20) Ə	gr.
94	11	3	0	19
85	9	1	1	15
91	10	7	2	14

Q. Repeat apothecaries' weight.

150.   CLOTH MEASURE.

(1)

yd.	(4) qr.	(4) na.
190	3	2
141	2	1
91	1	0
964	3	3
1388	2	2

(2)

yd.	(4) qr.	(4) na.
1000	2	3
976	3	3
1146	2	1
906	3	3

(3)

yd.	(4) qr.	(4) na.
9004	2	1
8094	3	2
16043	2	1
9076	2	3

Q. Repeat cloth measure.

151.   LONG MEASURE.

(1)

L.	(3) mi.	(8) fur.	(40) p.	(5½) yd.	(3) ft.
13	2	7	39	4	2
14	2	6	30	2	1
17	1	6	29	5	2
44	2	5	20	2	1
90	1	3	0	8	0

(2)

m.	(8) fur.	(40) p.	(5½) yd.	(3) ft.
100	7	30	5	2
94	3	21	2	1
95	2	19	5	1
43	7	37	4	2

(3)

p.	(5½) yd.	(3) ft.
19	4	2
47	3	2
105	5	1
96	3	2

Q. Repeat long measure.

152. LAND OR SQUARE MEASURE.

(1)

A.	(4) R.	(40) P.
94	3	39
160	2	17
90	2	35
864	3	24
1151	0	35

(2)

A.	(4) R.	(40) P.	(30½) sq. yd.
20	3	35	30
249	2	37	17
904	2	29	16
640	2	37	13

(3)

(9) sq. ft.	(144) sq. in.
8	140
7	64
5	109
8	96

(4)

M.	(640) A.	(4) R.	(40) P.	(30½) sq. yd.
1	527	3	29	30
14	294	2	30	24
16	350	1	37	19
84	579	3	39	29

Repeat land measure.

153. SOLID OR CUBIC MEASURE.

(1)

T.	(40) cu. ft.	(1728) cu. in.
4	39	1700
14	16	497
3	14	1497
22	31	238

(2)

cords.	(128) cu. ft.	(1728) cu. in.
49	120	1644
52	125	964
94	121	1535

(3)

cu. yds.	(27) cu. ft.	(1728) cu. in.
197	26	1500
946	24	1727
1000	25	1462

Q. Repeat solid measure.

154. LIQUID MEASURE.

(1)

hhd.	(2) bbl.	(31½) gl.	(4) qt.	(2) pt.
14	1	30	3	1
15	0	27	2	1
29	1	14	1	1
35	1	29	3	0
96	0	152	1	

(2)

T.	(2) P.	(2) hhd.	(2) bbl.	(31½) gl.
19	1	1	1	30
904	0	1	1	27
44	1	1	1	15
59	1	0	1	19

(3)

bbl.	(31½) gl.	(4) qt.	(2) pt.	(4) gi.
94	31	3	1	3
161	28	2	1	3
73	18	3	1	2
168	17	2	1	3

Q. Repeat liquid measure.

F 2

155. DRY MEASURE.

(1)			
Bush.	[4] pk.	(2) gl.	(4) qt.
19	3	1	3
25	2	1	3
23	3	1	3
100	2	1	2
99	3	1	3
270	1	0	2

(2)		
chald.	(36) bush.	(4) pk.
47	33	3
91	33	2
108	31	8
97	27	2
43	21	3

(3)				
bush.	(4) pk.	(2) gl.	(4) qt.	(2) pt.
14	3	1	1	1
94	2	1	3	1
109	3	1	3	1
86	2	1	3	1
88	3	1	2	0

Q. Repeat dry measure.

156. TIME.

(1)			
(12) yr.	(4) mo.	(7) wk.	d.
3	11	3	6
15	10	2	5
29	11	3	6
43	9	2	5
93	8	1	1

(2)				
yr.	(12) mo.	(4) wk.	(7) d.	(24) hr.
114	10	3	6	25
98	7	2	6	19
33	5	3	5	15
49	0	2	4	19

(3)					
yr.	(12) mo.	(4) w.	(7) d.	(24) hr.	(60) m.
187	3	3	6	21	59
91	11	0	5	20	57
900	7	3	4	23	49
43	6	2	1	17	30

Q. Repeat time table.

157. CIRCULAR MEASURE.

(1)		
o	(60) ′	(60) ″
47	59	59
30	29	47
50	36	20
129	6	6

2)		
o	(60) ′	(60) ″
109	57	23
41	58	57
19	47	59

(3)		
o	(60) ′	(60) ″
90	45	35
41	27	34
91	21	33

Q. Repeat circular measure.

SUBTRACTION OF COMPOUND NUMBERS.

158. SUBTRACTION of Compound Numbers consists in finding how much one compound number exceeds another.

Let it be required to subtract £17 19s. 9d. from £24 3s. 4d.

Setting like units under each other, we commence as in subtraction of simple numbers on the right. We cannot take 9 pence from 4 pence, but borrowing 1 shilling from the minuend of the next column, or what is the same, 12 pence, and adding 12 to 4,

OPERATION.

£	s.	d.
24	3	4
17	19	9

Ans. 6 3 7

we have 16 pence; from which if we take 9 pence, we have 7 pence left. Set down 7 under the column of pence. Carrying now 1 shilling to the subtrahend of the next column to make up for the 1 we borrowed from the minuend, makes 20 shillings. 20 from 3 we cannot, but borrowing 1 pound or 20 shillings from the pounds of the minuend, and adding them to 3 shillings, makes 23 shillings. 20 from 23 leaves 3. Set down 3 under the column of shillings. Carrying 1 to the subtrahend in the next column, 18 from 24 leaves 6. Set down 6. The remainder is £6 3s. 7d.

From which we deduce the following

RULE.

I. *Set like units under each other.*

II. *Commencing on the right, subtract the number of each kind of unit in the subtrahend from the corresponding number in the minuend, and set the remainders under their respective units.*

III. *Should the number of any one unit in the subtrahend exceed the corresponding number in the minuend, add as many units to the minuend as make 1 of the next higher unit, and then subtract as before, carrying 1 to the subtrahend of the next higher unit.*

Q. What is subtraction of compound numbers? How are the numbers set down? Where do you commence to subtract? Where are the remainders set down? Should the number in the subtrahend exceed the *corresponding* number over it in the minuend, how do you subtract? When you borrow, what do you carry? Repeat the rule.

EXAMPLES.

	(1)		
	$	cts.	m.
From	140	37	5
Take	99	50	9
Ans.	40	86	6

	(2)		
	$	cts.	
From	914	25	
Take	18	31¼	
Ans.	895	93¾	

	(3)		
	$	cts.	
From	1000	00	
Take	94	62½	
Ans.	905	37½	

	(4)		
	£	s.	d.
From	94	13	4½
Take	77	15	9¼
Ans.	16	17	7¼

	(5)		
	£	s.	d.
From	100	10	3¼
Take	91	12	4½
Ans.	8	17	10¾

	(6)			
	cwt.	qr.	lb.	oz.
From	9	3	1	12
Take	3	2	9	15
	6	0	19	13

	(7)			
	lb.	oz.	pwt.	gr.
From	24	4	10	3
Take	15	9	15	20

	(8)				
	℔	ℨ	ℨ	Ə	gr.
From	17	4	2	1	10
Take	10	9	1	1	15
	6	7	0	2	15

	(9)		
	yd.	qr.	na.
From	100	1	1
Take	87	3	2

	(10)		
	yd.	ft.	in.
From	47	2	2
Take	19	2	9
	27	2	5

	(11)		
	A.	R.	P.
From	19	2	19
Take	14	3	36

	(12)		
	cord.	cu. ft.	cu. in.
From	100	100	100
Take	50	119	1000

	(13)		
	bbls.	galls.	qt.
From	99	3	1
Take	70	19	2

	(14)		
	bush.	pk.	gal.
From	144	0	0
Take	19	1	1

	(15)				
	yr.	mo.	wk.	dy.	hr.
From	14	3	2	4	4
Take	10	10	3	5	5

(16) From 90° take 41' 45".

MULTIPLICATION OF COMPOUND NUMBERS.

159. MULTIPLICATION of Compound Numbers, like that of simple numbers, consists in repeating the multiplicand as many times as there are units in the multiplier.

Multiply £14 2s. 3d. by 2.

Placing the multiplier under the lowest unit of the multiplicand, we say, 2 times 3 pence are 6 pence; set down 6 under the pence: 2 times 2 shillings are 4 shillings; set down the 4 under the shillings: 2 times 4 are 8, 2 times 1 are 2. £28 4s. 6d. will be the answer.

OPERATION.

£	s.	d.
14	2	6
		2

Ans. 28 4 6

Again, multiply £17 8s. 9d. by 5.

Setting the numbers down as in the last example, we say 5 times 9 pence are 45 pence; but since 12 pence make 1 shilling, 45 pence contain $\frac{45}{12}$ shillings, or 3 shillings and 9 pence over: we set down the remainder 9 under the pence, and retain the 3 shillings to carry; 5 times 8 shillings are 40 shillings, and

OPERATION.

£	s.	d.
17	8	9
		5

Ans. 87 3 9

3 we carry are 43 shillings, but 20 shillings make 1 pound; there are then $\frac{43}{20}$ pounds in 43 shillings, that is, 2 pounds and 3 shillings over; set down 3 under the shillings, and retain the 2 pounds to carry; 5 times 7 are 35, and 2 we carry are 37; set down 7 and carry 3: 5 times 1 are 5 and 3 are 8. The answer is £87 3s. 9d.

Again, multiply £15 5s. 4d. by 24.

Since the multiplier in this case exceeds 12, and can be divided into the two factors 6 and 4, the operation is shortened by multiplying by these factors separately, as in multiplication of simple numbers (Art. 37).

OPERATION.

£	s.	d.
15	5	4
		6
91	12	0
		4

Ans. 366 8 0

Again, multiply £12 8s. 3½d. by 39.

Since the multiplier in this case cannot be divided into two entire factors, we take the number nearest to 39, which can be divided into factors, viz. 36=6×6, and multiply by each of these factors; then multiply the given number by the difference between the assumed number and the given multiplier, and add this product to that obtained by the multiplication by the two factors 6 and 6. The reason of this is plain, for as in multiplication the multiplicand has to be repeated as many times as there are units in the multiplier, the multiplication by the factors used only repeats the multiplicand 36 times;

1st OPERATION.

£	s.	d.
12	8	3¼
		6
74	9	9
		6
446	18	6
37	4	10½

Ans. 484　3　4¼

2d OPERATION.

£	s.	d.
12	8	3¼
		3
37	4	10½

and then multiplying it by 3, it was repeated 3 times more, that is, 36+3=39 times.

From these illustrations we deduce the following

RULE.

I. *Set the multiplier under the lowest unit of the multiplicand, and draw a line beneath.*

II. *Multiply the number on the right by the multiplier and divide this product by as many units as make one of the next higher units. Set the remainder, if any, under the number multiplied, and carry the quotient into the next product. Multiply the number of the next higher units by the multiplier, and to this product add the number carried. Divide this sum by as many units as make one of the next higher units, and proceed in this way until all the units are multiplied, setting down the entire product when the highest unit is multiplied.*

III. *When the multiplier exceeds 12, and can be divided into factors, multiply the given number by one of these factors, and this product by the second factor, and so on until all the factors are used: the last product will be the required product.*

IV. *When the multiplier exceeds* 12, *and cannot be divided into entire factors, take the nearest less number to the given multiplier which can be divided into factors, and multiply by them as before ; then multiply the given multiplicand by the difference between the given multiplier and the assumed number, and add this product to the last product : the sum will be the required product.*

The proof is the same as in multiplication of simple numbers.

Q. What is multiplication of compound numbers? How are the numbers set down? Where do you commence to multiply? Do you set down the whole product? What do you set down? What do you carry? When the multiplier exceeds 12, and can be divided into factors, how may you multiply? When the multiplier exceeds 12, and cannot be divided into factors? What is the reason of this? What is the rule for multiplication of compound numbers? What is the proof?

EXAMPLES.

	(1)			(2)		(3)	
	$	cts.	m.	$	cts.	$	cts.
Multiply	14	25	4	413	37½	240	31¼
by			5		9		10
	71	27	0	3720	37½	2403	12½

4. Multiply $291.68 by 25.

OPERATION.

We reduce in this example the multiplicand to cents, then multiply as in simple numbers, pointing off two figures to the right to bring the product to dollars.

cts.
29168
25

145840
58336

$7292.00 cts.

5. Multiply £24 2s. 9d. by 32.   Ans. £772 8s.
6. Multiply £109 12s. 4½d. by 96.   Ans. £10523 8s.
7. Multiply 9 cwt. 2 qr. 27 lb. by 18. Ans. 175 cwt. 1 qr. 10 lb.
8. Multiply 14 A. 3 R. 27 P. by 47.
9. Multiply 10 cords 4 cu. ft. 12 cu. in. by 49.
10. Multiply 41 gals. 2 qt. 1 pt. by 63.
11. Multiply 10 mo. 2 wk. 6 dy. 14 h. by 108.
12. Multiply 70° 4' 22" by 180.

DIVISION OF COMPOUND NUMBERS.

160. LET it be required to divide £18 8s. 4d. by 2.

We place the divisor on the left of the dividend, as in division of simple numbers; draw a curve line to separate them, and then a straight line beneath: 2 in 18 pounds goes 9 pounds; set down 9 under pounds: 2 in 8 shil-

```
    OPERATION.
     £   s.  d.
2 ) 18   8   4
   ─────────────
Ans. 9   4   2
```

lings goes 4 shillings; set down the 4 under shillings: 2 in 4 pence goes 2 pence; set down 2 under pence: the quotient is £9 4s. 2d.

Again, divide £17 5s. 4d. by 4.

In this example 4 in 17 pounds goes 4 pounds, and 1 pound over; set down 4 under pounds. But since the 1 pound equals 20 shillings, and has not been divided, we add 20 to 5 shillings, which make 25 shillings; 4

```
    OPERATION.
     £   s.  d.
4 ) 17   5   4
   ─────────────
Ans. 4   6   4
```

in 25 shillings goes 6 shillings, and 1 shilling over; 1 shilling is equal to 12 pence, and adding the 12 pence to the 4 pence makes 16 pence, which divided by 4 gives 4 pence. The answer is £4 6s. 4d.

Again, divide £65 14s. 8d. by 16.

The divisor in this example being greater than 12, we arrange the numbers as in division of simple numbers. Dividing the 65 pounds by 16, we get 4 pounds for the quotient, and 1 pound for a remainder. Multiplying this remainder by 20 to bring it to shillings (Art. 118), and adding to them the 14 shillings which have not been divided, gives us 34 shillings. Taking this number as a new *dividend*, and dividing again by 16, we get 2 shillings for the quotient and 2 shillings

```
       OPERATION.
      £    s.  d.
16 ) 65   14   8 ( 4 pounds.
     64
    ────
      1
     20
    ────
  16 ) 34 ( 2 shillings.
       32
      ────
        2
       12
      ────
   16 ) 32 ( 2 pence.
        32

   Ans. £4 2s. 2d.
```

over. Multiplying this remainder by 12 to bring the shillings to pence, and adding in the 8 pence, gives us 32 pence; which, divided by 16, gives 2 pence and no remainder. The answer is £4 2s. 2d.

From these illustrations we deduce the following

RULE.

I. *Arrange the numbers as in division of simple numbers*

II. *Divide the number corresponding to the highest unit by the divisor : the quotient will be the number corresponding to the highest unit in the quotient.*

III. *Multiply the remainder by as many of the next lower units as make one of the higher, and to this product add the number of the next lower units in the dividend ; divide as before ; the quotient will be the number corresponding to the next lower unit in the quotient.*

IV. *Proceed in this manner through all the different units, and set the last remainder, if there be any, as in division of simple numbers.*

The proof is the same as for simple numbers.

Q. How are the numbers arranged for division of compound numbers? Where do you commence to divide? Suppose there is a remainder, how do you proceed? After multiplying this remainder, what do you add to it? What do you do with the last remainder? What is the proof for simple numbers?

EXAMPLES.

1. Divide $104 37½ cents by 5.

The ½ cent is brought to decimals, which is .5, and then divide as in simple numbers, pointing off for dollars and cents.

OPERATION.

5) 104.37.5

Ans. $20.87.5

2. Divide $1000 by 33.   Ans. $30.30 cts.+
3. Divide $889.37½ by 84.   Ans. $10.58 cts. 7m.+
4. Divide £109 14s. 3½d. by 11. Ans. £9 19s. 5½d.
5. Divide 10 lb. 10 oz. by 17.   Ans.
6. Divide 100 cwt. by 16.   Ans.
7. Divide 16 bu. 1 pk. by 12.   Ans.
8. Divide 10 A. 1 R. 3 P. by 47.   Ans.

161. When the divisor can be separated into factors, we may divide by each of these factors separately, as in division of simple numbers (Art. 56).

Thus, divide 14 cwt. 2 qrs. 20 lbs. by 24.

OPERATION.

The factors of 24 being 6 and 4, we first divide by 6, and then divide the quotient by 4. The answer is 0 cwt. 2 qr. 12¾ lb.

$$
\begin{array}{r}
\text{cwt.} \quad \text{qr.} \quad \text{lb.} \\
6\,)\,14 \quad\ 2 \quad 20 \\
\hline
4\,)\,2 \quad\ 1 \quad 22 \\
\hline
0 \quad\ 2 \quad 12\tfrac{3}{4}
\end{array}
$$

Q. How may you perform the division when the divisor can be separated into its factors?

9. Divide 12 mos. 2 wk. 2 dy. by 16.   Ans. 0 m. 3 w. 1 d.

10. Divide 180° by 144.

11. Divide 116 m. 2 fur. by 28.

12. Divide 34 bbls. 21 galls. by 42.

REDUCTION OF VULGAR FRACTIONS

WHICH ARE EXPRESSED IN TERMS OF DIFFERENT UNITS.

162. THE Vulgar Fractions which have been heretofore considered, have expressed parts of the *same kind* of unit. But in arithmetical operations, fractions are frequently presented for addition, &c., in which parts of *different* units are considered.

Thus, it may be required to add ⅓ of a pound to ¼ of a shilling; and it is evident that before this addition can be effected, the ⅓ of a pound must be reduced to the same unit as ¼ of a shilling, or the reverse, the fraction of the shilling must be brought to the same unit as that of the pound.

Thus, it appears that to prepare fractions of different units for the operations of addition, &c., they must first be reduced to the same unit, and this reduction may be effected in two ways:

1st. By reducing a fraction of a higher unit to a fraction of a lower.

2dly. By reducing a fraction of a lower unit to a fraction of a higher.

Q. What kind of fractions have been previously considered ? What is necessary before you can add fractions of different units? Can you add before reduction $\frac{1}{3}$ of a pound to $\frac{1}{4}$ of a shilling ? Why not ? What is necessary before they can be added ? In how many ways can this reduction be effected ? What are they ?

CASE I.

163. *To reduce a fraction of a higher unit to a fraction of a lower.*

Reduce $\frac{1}{7}$ of a £ to the fraction of a penny.

Since £1 is equal to 20 shillings, $\frac{1}{7}$ of a pound is equal to $\frac{1}{7} \times 20 = \frac{20}{7}$ shillings. But 1 shilling is equal to 12 pence ; $\frac{20}{7}$ shillings are consequently equal to $\frac{20}{7} \times 12 = \frac{240}{7} d.$

OPERATION.

$\frac{1}{7} \times 20 \times 12 = \frac{240}{7} d.$

Again, reduce $\frac{1}{3}$ of a $ to the fraction of a mill.

Here $1 being equal to 100 cents, $\frac{1}{3}$ of a $ is equal to $\frac{1}{3} \times$ 100 cents ; and 1 cent being equal to 10 mills, $\frac{1}{3} \times 100$ cents will be equal to $\frac{1}{3} \times 100 \times 10$ mills $= \frac{1000}{3}$ mills.

OPERATION.

$\frac{1}{3} \times 100 \times 10 = \frac{1000}{3}$ m.

From these illustrations we deduce the following

RULE.

Multiply the given fraction by as many of the next lower units as make one of the higher, and so on through all the different units, until you reach the unit to which the fraction is to be reduced.

Q. How many shillings make 1 pound ? How many shillings is $\frac{1}{7}$ of a £ equal to? How many pence make 1 shilling ? How many pence make $\frac{20}{7}$ of a shilling ? How many cents will $\frac{1}{3}$ of a $ be equal to? How many mills will $\frac{100}{3}$ cents be equal to? What is the rule to reduce a fraction of a higher unit to a fraction of a lower ?

EXAMPLES.

1. Reduce $\frac{3}{5}$ of a £ to the fraction of a penny.

                Ans. $7\frac{2}{5}0$ pence.

2. Reduce $\frac{1}{13}$ of a £ to the fraction of a penny.

                Ans. $^2\frac{4}{13}^0$ pence.

3. Reduce $\frac{17}{20}$ of a $ to the fraction of a mill.

                Ans. $^{17}\frac{0}{20}^0$ mills.

4. Reduce $\frac{3}{11}$ of an eagle to the fraction of a cent.

                Ans. $^3\frac{00}{11}^0$ cents.

5. Reduce $\frac{1}{8}$ of a cwt. to the fraction of a lb.

                Ans. $1\frac{1}{4}^2$ lbs.

6. Reduce $\frac{4}{5}$ of a mile to the fraction of a foot.

                Ans. $^{21}\frac{1}{5}^{20}$ ft.

7. Reduce $\frac{17}{19}$ of a barrel to the fraction of a quart.

                Ans.

8. Reduce $\frac{14}{15}$ of a lb. Troy to the fraction of an ounce.

                Ans.

9. Reduce $\frac{21}{19}$ of a bushel to the fraction of a gallon.

                Ans.

10. Reduce $\frac{114}{119}$ of a day to the fraction of a minute.

                Ans.

11. Reduce $\frac{125}{242}$ of a degree to the fraction of a second.

                Ans.

12. Reduce $\frac{13}{20}$ of a circumference to the fraction of a degree.   Ans.

164. It follows from the last article that we may readily *reduce a fraction of a given unit to whole numbers of a lower unit; that is, find its value in these terms.*

Thus, to find the value of $\frac{4}{5}$ of a £ in whole numbers of a lower unit,

We multiply the given frac-   OPERATION.
tion of the pound by 20 to $\frac{4}{5} \times 20 = \frac{80}{5} = 16$ shillings.
bring it to shillings, which is
equal to $\frac{80}{5}$ shillings, the value of which is 16 shillings.

Again, find the value of $\frac{5}{6}$ of a £ in whole numbers of lower units.

OPERATION.

$$\begin{array}{r} 5 \text{ pounds.} \\ 20 \\ \hline 6\,)\,100 \text{ shillings.} \\ \hline \text{shillings } 16\text{—}4 \text{ sh. over} \\ 12 \\ \hline 6\,)\,48 \\ \hline 8 \text{ pence.} \end{array}$$

Ans. 16*s*. 8*d*.

We first multiply by 20 to bring the pounds to shillings, which produces $\frac{100}{6}$ shillings; or 16 shillings and 4 over. We then multiply the $\frac{4}{6}$ shillings by 12 to bring to pence, which gives $\frac{48}{6}$ pence, or 8 pence. The answer then is 16*s*. 8*d*.

We have therefore the following

RULE.

Multiply the given fraction by as many of the next lower units as make one of the higher, and divide by the denominator of the fraction. Multiply the remainder, if there be any, by as many of the next lower units still, and reduce as before, continuing this process through all the different units as long as there is a remainder. The various quotients connected together will be the answer.

Q. How may you find the value of a fraction of a unit in whole numbers of a lower unit? By what do you multiply? By what do you divide? If there be a remainder, how do you proceed? How far do you carry the process?

EXAMPLES.

1. Find the value of $\frac{4}{5}$ of a $. Ans. 80 cents.
2. Find the value of $\frac{1}{6}$ of a $. Ans. 16+ cents.
3. Find the value of $\frac{11}{12}$ of a cwt.
   Ans. 3 qr. 18 lb. 10 oz. 10 dr.+
4. Find the value of $\frac{10}{9}$ of a mile.
   Ans. 8 fur. 35 R. 3 yds.+
5. Find the value of $\frac{2}{5}$ of a barrel.
6. Find the value of $\frac{18}{19}$ of a bushel.
7. Find the value of $\frac{14}{15}$ of a day.
8. *Find the value of* $\frac{3}{5}$ of a degree.

CASE II.

165. *To reduce a fraction of a lower unit to a fraction of a higher.*

Reduce $\frac{2}{3}$ of a penny to the fraction of a pound.

Since 12 pence make 1 shilling, 1 penny is equal to $\frac{1}{12}$ shilling, 2 pence to $\frac{2}{12}$ shillings, &c., the value

OPERATION.

$$\frac{1}{2\times12\times20}=\frac{1}{480}\;£.$$

of the pence in terms of the shilling being obtained by dividing the pence by 12, the number of pence in a shilling. Hence $\frac{1}{2}$ of a penny $=\frac{1}{2}\div12=\frac{1}{2\times12}$ (Art. 67)$=\frac{1}{24}$ shillings. But 1 shilling is equal to $\frac{1}{20}$ pounds; hence $\frac{1}{24}$ shilling $=\frac{1}{24}\div20=\frac{1}{24\times20}=\frac{1}{480}\;£.$

We therefore have the following

RULE.

Divide the given fraction by as many units as make 1 of the next higher units, and proceed in this way through all the different units, until you reach the unit to which the reduction has to be made.

Q. What part of a shilling is 1 penny? Are 2 pennies? Is $\frac{1}{2}$ penny? How is the value of pence found in terms of a shilling? What part of a pound is 1 shilling? Is $\frac{1}{24}$ of a shilling? How is the value ascertained? What is the rule for this reduction?

EXAMPLES.

1. Reduce $\frac{1}{4}$ of a penny to the fraction of a pound.

Ans. $\frac{1}{720}\;£.$

2. Reduce $\frac{5}{7}$ of a penny to the fraction of a pound.

Ans. $\frac{1}{336}\;£.$

3. Reduce $\frac{3}{4}$ of a mill to the fraction of a dollar.

Ans. $\frac{3}{4000}\;$.

4. Reduce $\frac{4}{5}$ of a cent to the fraction of an eagle.

Ans. $\frac{1}{1250}$ E.

5. Reduce $\frac{4}{9}$ of an ounce to the fraction of a cwt.

6. Reduce $\frac{2}{5}$ of a pint to the fraction of a hhd.

7. Reduce $\frac{4}{9}$ of an hour to the fraction of a month.

8. Reduce $\frac{4}{9}$ of a foot to the fraction of a mile.

Reduce $\frac{4}{11}$ of a second to the fraction of a degree.

166. *A whole number may by the same principle be reduced to a fraction of a higher unit.*

Thus, we may reduce 3*s.* 7*d.* to the fraction of a £.

In this example we reduce the shillings and pence to pence (by Art. 118), multiplying by 12, which gives us 43 pence. But 43 pence are equal to $\frac{43}{20}$ shillings (Art. 165), or $\frac{43}{12\times20} = \frac{43}{240}$ £.

OPERATION.

s.	*d.*
3	7
12	

43 pence.

Ans. 43*d.* = $\frac{43}{240}$ £

Hence we have the following

RULE.

Reduce the given number to the terms of the lowest unit expressed in it, and place it as a numerator over as many of the same units as make 1 of the units to which the number is to be reduced as a denominator. Then reduce this fraction to its lowest terms.

Q. May whole numbers be reduced to a fraction of a higher unit? How are shillings and pence brought to pence? How are pence expressed in fractions of a pound? Of a shilling? What is the rule for this reduction?

EXAMPLES.

1. What part of a pound is 2*s.* 8*d.*?   Ans. $\frac{4}{30}$ £.
2. What part of a $ is 33 cts. 2 m.?   Ans. $\frac{332}{1000}$ $.
3. What part of an eagle is $1 34 cts.?   Ans. $\frac{134}{1000}$ E.
4. What part of a bushel is 3 pk. 2 gals. 1 qt.?
5. What part of a mile is 17 yards 2 ft.?
6. What part of a day is 13 hrs. 17 min.?
7. What part of an acre is 2 R. 12 P.?
8. What part of a degree is 35′ 41″?

REDUCTION OF DECIMAL FRACTIONS.

167. To reduce a compound number to the decimal of a higher unit.

Thus, reduce 14s. 6d. to the decimal of a £.

Since 1 penny is $\frac{1}{12}$ of a shilling, 6 pence is $\frac{6}{12}$ of a shilling, which reduced to a decimal (Art. 97), is .5s. The given number then becomes 14.5 shillings. Dividing this by 20 will bring it to pounds. Hence .725 £ is the decimal required.

OPERATION.

12) 6.0 pence.
    ————
     .5 shillings.
20) 14.5 shillings.
    ————
Ans. .725 £.

We may therefore deduce the following

RULE.

Divide the lowest number corresponding to the lowest unit expressed, by as many as make one of the next higher units. Place the quotient as the decimal part of the number corresponding to the next higher unit. Divide again by as many as make 1 of the next higher unit still, and continue in this way until you have reached the unit to which the reduction has to be made; the last quotient will be the answer.

Q. How do you reduce a compound number to a decimal of a higher unit? By what would you divide pence to bring them to the decimal of a shilling? The decimal of a pound? By what would you divide shillings to bring them to the decimal of a pound?

EXAMPLES.

1. Reduce 9 pence to the decimal of a £.

Ans. £.0375.

2. Reduce 1 pwt. to the decimal of a pound troy.

Ans. .004166 + lb.

3. Reduce 7 drachms to the decimal of a pound avoirdupois.

Ans. .02734375 lb.

4. Reduce 26 pence to the decimal of a £.

Ans. £.10833 + .

5. Reduce .056 poles to the decimal of an acre.

Ans. .00035 A.

6. Reduce 17*s.* 9¾*d.* to the decimal of a pound.

Ans. £.890625.

7. Reduce £19 17*s.* 3¼*d.*   Ans. £19.863+.

8. Reduce 15*s.* 6*d.* to the decimal of a £.

Ans. £.775.

9. Reduce 3 R. 35 P. to the decimal of an acre.

Ans.   A.

10. Reduce 23 hrs. 49 m. to the decimal of a day.

Ans.   d.

11. Reduce 5 oz. 12 pwts. 16 grs. to the decimal of a lb.

Ans. .46944+lb.

12. Reduce 4° 37′ 21″ to the decimal of a degree.

Ans.

168. *To reduce decimals of a higher unit to whole numbers of a lower unit.*

Thus, reduce £.275 to its proper value in whole numbers.

OPERATION.

The principle of this rule corresponds with that of Art. 116. The pounds are multiplied by 20 to bring them to shillings, pointing off as in multiplication of decimals (Art. 94). The shillings are then multiplied by 12 to bring them to pence. The answer is 5*s.* 6*d.*

£.275
20
———
5.500 shillings.
12
———
6.000 pence.

Ans. 5*s.* 6*d.*

RULE.

Multiply the given decimal by as many of the next lower units as make 1 of the given units, and point off from the right of the product as many decimals as there are in the multiplicand. Multiply the decimal part of this product by as many of the next lower units still as make 1 of the units corresponding to the new multiplicand, and cut off as before.

G

Continue this operation through all the different units. The whole numbers in each separate product connected together will be the answer.

Q. How may a decimal of a given unit be reduced to whole numbers of a lower unit? How may .5 of a shilling be brought to pence? How many pence will it be equal to? .2 of a £ brought to shillings? .5 of a $ to cents? .5 of a ct. brought to mills? What is the value of .75 of a $? Of .4 of a $?

EXAMPLES.

1. What is the value of .333 dollars?   Ans. $3.33 cts.
2. What is the value of £.775?   Ans. 15s. 6d.
3. What is the value of £.4765?   Ans. 9s. 6¼d. +
4. What is the value of .3376 lb. avoirdupois?

    Ans. 5 oz. 6 dr. +

5. What is the value of .104376 bush.?
6. What is the value of .946 yds.?
7. What is the value of .04967 A.?
8. What is the value of .0046987 miles?
9. What is the value of .00946 cwt.?

PROPORTION.

169. Two quantities of the same kind may be compared together in two ways:

1st. *By considering how much one exceeds the other, and their relation will be shown by their difference.*

Thus, by comparing the numbers 7 and 5 with respect to their difference, we find that 7 exceeds 5 by 2.

2dly. *By considering how many times one contains the other, and their relation will be shown by their quotient.*

Thus, by comparing 12 and 3 to ascertain how many times 12 contains 3, the quotient $\frac{12}{3} = 4$ shows this relation.

Q. In how many ways may quantities of the same kind be compared? What is the first method? What shows the relation between the two quantities in this case? Give an example. What is the second method? What shows the relation between the two quantities in this case? Give an example.

170. RATIO *is the relation which one quantity bears to another of the same kind.*

When the quantities are compared by considering *how much* one exceeds the other, the ratio, which is expressed by their difference, is called their *arithmetical ratio*, or simply their *difference.*

Thus, the arithmetical ratio of 7 to 5 is 7—5=2; of 10 to 8 is 10—8=2.

When the quantities are compared by considering *how many times* one contains the other, the ratio is called their *geometrical ratio*, and is expressed by dividing the first by the second.

Thus, the geometrical ratio of 12 to 3 is $\frac{12}{3}$ or 4; that of 3 to 12 is $\frac{3}{12}$. A geometrical ratio may therefore be entire or fractional.

Q. What is meant by ratio? What is an arithmetical ratio? What is the arithmetical ratio of 7 to 5? 5 to 3? 3 to 1? What is a geometrical ratio? What is it expressed by? What is the geometrical ratio of 6 to 3? Of 8 to 4? Of 10 to 5? Of 3 to 4? Of 4 to 3? Is a geometrical ratio ever fractional? Is it in the ratio of 4 to 2? Of 2 to 4?

171. The numbers which enter into a ratio are called its *terms.* Every ratio then has *two* terms. The first term of a ratio is called the *antecedent;* the one with which it is compared the *consequent.*

In the arithmetical ratio 5—2, 5 is the *antecedent*, 2 the *consequent.*

In the geometrical ratio $\frac{5}{4}$, 5 is the *antecedent*, 4 the *consequent.*

Q. What are the numbers called which enter into a ratio? Which is the antecedent? Which the consequent? In the ratio 5—2, which is the antecedent? Which the consequent? In the ratio $\frac{5}{4}$, which is the antecedent? Which the consequent?

172. An arithmetical ratio is not changed in value when its terms are increased or diminished by the same number, because their difference, which expresses this ratio, remains the same.

Thus, the arithmetical ratio of 12 to 5 is the same as that of 14 to 7, and of 10 to 3, although in the first case the terms were increased by the number 2, and in the second diminished by the same number.

173. A geometrical ratio is not changed in value by multiplying or dividing its terms by the same number. For this ratio is expressed by placing the first term over the second, as a fraction, and a fraction is not changed in value by multiplying or dividing its terms by the same number (Art. 69 and 70).

Thus, the geometrical ratio of 4 to 2 is the same as that of 8 to 4, or of 2 to 1, since $\frac{4}{2}=2$, $\frac{8}{4}=2$, $\frac{2}{1}=2$.

174. Four numbers are said to be in *arithmetical proportion,* when the difference between the first and second is equal to the difference between the third and fourth. This is also called a proportion by *differences.*

Thus, the four numbers 4, 2, 10, 8, are in arithmetical proportion, since $4-2=10-8$.

Four numbers are said to be in *geometrical proportion,* when the quotient arising from the division of the first by the second is equal to that arising fron the division of the third by the fourth. This is also called a proportion by *quotients.*

Thus, the four numbers, 4, 2, 16 and 8, are in geometrical proportion, since $\frac{4}{2}=\frac{16}{8}$.

PROPORTION, *therefore, is an equality of two ratios.*

175. There are *four* terms in every proportion. The first and fourth terms are called the *extremes;* the second and third the *means.*

An arithmetical proportion is expressed by writing one dot between the first and second terms; two dots between the second and third terms; and one dot between the third and fourth terms.

Thus, **4 . 2 : 10 . 8,**

which is read

4 *is to* 2 *as* 10 *is to* 8.

4 and 8 are the *extreme* terms; 2 and 10 the *means.*

A geometrical proportion is expressed by *doubling* the number of dots used in an arithmetical proportion. Thus,

4 : 2 :: 16 : 8,

which is read

4 *is to* 2 *as* 16 *is to* 8;

4 and 8 are the *extremes;* 2 and 16 the *means.*

Q. How many terms in a proportion? Which are the extremes? The means? How is an arithmetical proportion expressed? Write one. Which are the extremes? Which the means? How is a geometrical proportion expressed? How read? Write one. Which are the extremes? Which the means?

176. The first and second terms of an arithmetical proportion may be made equal, and also the third and fourth terms, by adding, in each case, to the smaller term, the ratio, or by subtracting the ratio from the greater terms.

Thus, in the proportion

4 . 2 : 10 . 8,

the ratio is 2, and by adding it to 2 and 8, we have

4 . 4 : 10 . 10;

and by subtracting it from 4 and 10, we have

2 . 2 : 8 . 8.

Q. How may the first two terms and the last two terms of an arithmetical proportion be made equal?

177. The first and second terms of a geometrical proportion may be made equal, and also the third and fourth, by multiplying the smaller or dividing the larger terms in either case by the ratio.

Thus, in the proportion

4 : 2 :: 16 : 8,

we have, by multiplying the second and fourth terms by the ratio 2,

$$4 : 4 :: 16 : 16;$$

and by dividing the first and third terms by the same ratio, we have

$$2 : 2 :: 8 : 8.$$

Q. How may the first two and last two terms of a geometrical ratio be made equal?

178. *The fundamental property of an arithmetical proportion is, that the sum of the extremes is equal to the sum of the means.*

Thus, in the proportion

$$3 . 7 : 8 . 12$$
$$3 + 12 = 7 + 8;$$

and this is true for every similar proportion. For, if the first two terms were equal to each other, and also the last two, as in the proportion

$$7 . 7 : 12 . 12,$$

it is evident the sum of the extremes must be equal to the sum of the means, since we have

$$7 + 12 = 7 + 12;$$

and (by Art. 176), this reduction can always be effected.

Q. What is the fundamental property of an arithmetical proportion?

179. *The fundamental property of a geometrical proportion is, that the product of the extremes is equal to the product of the means.*

Thus, in the proportion

$$4 : 2 :: 16 : 8,$$

we have

$$4 \times 8 = 2 \times 16, \text{ or } 32 = 32;$$

and this is true in every such proportion. For, if the first two and last two terms were equal to each other, as in the proportion

$$4 : 4 :: 16 : 16,$$

it is evident the product of the extremes must be equal to the product of the means, since we have

$$4 \times 16 = 4 \times 16;$$

and (by Art. 177), this reduction can always be effected.

Q. What is the fundamental property of a geometrical proportion?

180. It follows from this fundamental property of a geometrical proportion, that *knowing the first three terms of a proportion, we may determine the fourth by multiplying the second and third terms together, and dividing by the first.*

For, by Art. 178, the product of the means is equal to the product of the first term by the fourth. The fourth term must therefore be equal to the product of the means divided by the first term.

Thus, to find the fourth term in the proportion

$$3 : 8 :: 12 : \text{4th term,}$$

we have 4th term $= \frac{12 \times 8}{3} = \frac{96}{3} = 32$;

the proportion then becomes

$$3 : 8 :: 12 : 32$$

in which we have $3 \times 32 = 8 \times 12$;

$$\text{or } 96 = 96,$$

as in Art. 179.

EXAMPLES.

1. The first three terms of a geometrical proportion are 1, 2, 3. What is the fourth term?   Ans. 6.
2. The first three terms of a geometrical proportion are 3, 6, 9. What is the fourth term?   Ans. 18.
3. The first three terms of a geometrical proportion are 1, 2, 2. What is the fourth term?   Ans. 4.
4. The first three terms of a geometrical proportion are 9, 2, 1. What is the fourth term?   Ans. $\frac{2}{9}$.

Q. How do you find the fourth term of a geometrical proportion of which you know the first three terms?

SINGLE RULE OF THREE.

181. *The* SINGLE RULE OF THREE *embraces a class of questions in which three numbers in geometrical proportion are given to find a fourth.*

Thus, if 25 yards of cloth cost $60, what will 84 yards cost?

In this example, three numbers are given: the 25 yards of cloth, their cost, $60, and the 84 yards of cloth; and it is required to find the fourth number, which is the cost of the 84 yards of cloth.

Now, since 25 yards of cloth cost $60, it is plain, 1, 2, 3, &c. times more cloth must cost 1, 2, 3, &c. times more dollars. Hence, the 25 yards and the 84 yards must bear the same relation to each other that their respective prices do. The four numbers are therefore in geometrical proportion, and to find the fourth term, we have only to apply the rule in Art. 180.

OPERATION.

yds. $ yds.
25 : 60 : : 84 : Ans.

The proportion being expressed by the signs in Art. 175, we multiply the second and third terms together, and divide by the first, and we find $201.60 the fourth term required.

$$60$$
$$\overline{}$$
$$25\,)\,5040.00\,(\$201.60$$
$$50$$
$$\overline{}$$
$$40$$
$$25$$
$$\overline{}$$
$$150$$
$$150$$
$$\overline{}$$
$$\ldots 0$$

If this answer be correct, the first term multiplied by it must be equal to the product of the means (Art. 179).

$$25 \times 201.60 = 5040.$$
$$60 \times 84 = 5040.$$

The answer is correct.

182. Let us take another example.

If 20 men can dig a certain ditch in 15 days, how many days will it take 30 men to dig it?

In this example, three numbers are again given, viz: 30 men, 20 men, and 15 days; and a fourth number is required. But if 20 men can dig the ditch in 15 days, it is plain that 2, 3, 4, &c. *times more* men would dig the same ditch in 2, 3, 4, &c. *fewer* days. The number of men, 20, will then be contained in the number 30, as many times as the number of days employed by the 30 men is in the 15 days employed by the 20 men.

OPERATION.

men. men. days.

$$30 : 20 :: 15 : Ans.$$

$$20$$

$$30 \,)\, 300$$

Ans. 10 days.

PROOF.

$$30 \times 10 = 300$$
$$20 \times 15 = 300$$

The four numbers are therefore in proportion, and the operation shows that the fourth term is 10 days.

The product of the extremes being still equal to the product of the means, the answer is correct.

Q. What class of questions does the single rule of three embrace? In the first example, which are the three given numbers? What is the required number? If 1 yard of cloth cost $2, what will 2 yards cost? 3 yards? 4 yards? If 2 yards cost $4, what will 3 yards cost? 4 yards? 5 yards? 6 yards? Does more cloth require more or less money? What relation does the number of yards bear to the prices in each case? The four numbers then form what? In what manner may the fourth term be found? How do you know when you have the correct answer? In the second example, which are the three given numbers? What number is required? If 2 men can dig a ditch in 20 days, how many days would it take 4 men to dig it? 8 men? Do more men then require more time or less time? What relation does the number of days bear to the number of men? The four numbers form what? How is the fourth term found? How do you know that the answer is correct?

183. The two examples just explained exhibit the two classes of questions which are embraced in the *Single Rule of Three.*

In both cases there are four numbers considered, two of one kind of quantity, and the other two of another kind of quantity.

In the first case, there is a direct relation between the number of yards of cloth and their corresponding prices

G 2

the *greater* number of yards requiring the *greater* number of dollars, and the *less* number of yards the *less* number of dollars.

A proportion in which the corresponding terms are so related that *more* requires *more*, or *less* requires *less*, is called a *direct proportion*.

In the second example, *more* men perform the same work in *less* days, and *less* men in *more* days. The relation here between the men and their corresponding number of days is consequently *inverted*. A proportion in which the corresponding terms are so related, that *more* requires *less*, or *less* requires *more*, is called an *inverse* proportion.

From what has been said, we may form for both of these classes of questions the following general

RULE.

Place the number which is of the same kind as the answer sought in the third term.

Consider whether, from the nature of the question, the answer is greater or less than the third term. If it is greater, place the least of the remaining numbers in the first term, and the greater in the second term. But if the answer be less than the third term, place the greater of the remaining numbers in the first term, and the less in the second.

Multiply the second and third terms together, and divide by the first; the quotient will be the fourth term or answer sought.

PROOF.

Multiply the extremes together, and also the means; if the answer be correct, the two products will be equal.

Q. How many kinds of questions are there in the single rule of three? How many quantities are considered in each case? What kind of quantities are they? In the first example, what kind of relation exists? Explain this. What is this proportion called? Why *direct?* In the second example, what kind of proportion? Why inverse? Are both of these cases worked by the same rule? What is this rule? What number is placed in the third term? Where are the two other given numbers placed? When the answer sought is greater than the third term, which is the second term? When the answer is less than the third term, which is the second term? How is the answer found? How do you know when the work is correct? What is the proof?

1. If 2 cwt. of sugar cost £4 13s. 4d., what will 20 cwt. cost?

In this example, £4 13s. 4d. is the third term, since it is of the same kind as the answer sought. As 20 cwt. of sugar will cost more than 2 cwt., we place 20 cwt. in the second term, and 2 cwt. in the first term. We then reduce the £4 13s. 4d. to pence by the rule for reduction descending; and then, to find the answer, multiply 1120 pence by 20, the second term, and divide by the first term. The quotient 11200 is the answer in pence; which, reduced to pounds, shillings, and pence, becomes £46 13s. 4d.

OPERATION.

```
cwt. cwt.      £  s. d.
 2 : 20  :: 4 13 4 : Ans.
               20
              ――
               93
               12
              ――
             1120
               20
             ――――
          2 ) 22400
             ――――
         12 ) 11200 pence.
             ――――
         2,0 ) 93,3―4 pence.
             ――――
             £46―13 sh.

             £   s.  d.
      Ans.   46  13  4
```

Instead of reducing the compound number to pence, we could have multiplied it at once by 20 or its factors 4 and 5, and then divided by 2, as follows:

OPERATION.

```
cwt. cwt.      £  s. d.
 2 : 20  :: 4 13 4 : Ans.
               4
             ――――
            18 13 4
               5
             ――――
          2 ) 93  6 8
             ――――
       Ans. 46 13 4
```

By this operation, we obtain the same answer. As the numbers are, however, often large, the reduction is most generally necessary.

2. If 3 cwt. of tobacco cost $115, what will 15 cwt. 3 qrs. 16 lb. cost?

OPERATION.

cwt.		cwt.	qr.	lb.	$
3	:	15	3	16	:: 115 : Ans.
4		4			

12	63
28	28

336 lb. : 1780 lb. :: $115 : Ans.

We first reduce the 1st and 2d terms to pounds, and then multiply the 2d and 3d terms together, and divide by the 1st as before. The answer is 609.22\frac{2}{28}$ cts.

336 lb. : 1780 lb. :: $115 : Ans.

$$115$$

$$8900$$
$$1780$$
$$1780$$

336) 204700 (609.22\frac{2}{28}$ Ans.
$$2016$$

$$3100$$
$$3024$$

$$760$$
$$672$$

$$880$$
$$672$$

$$\frac{108}{336} = \frac{9}{28}$$

3. If 100 barrels of flour will support 30 men 150 days, how long would they subsist 350 men?

OPERATION.

men.		men.		days.	
350	:	30	::	150	: Ans.

$$30$$

The answer being days, the third term is days; and since the answer is less than the third term, the second term is less than the first. The result gives 12$\frac{6}{7}$ days for the answer.

350) 4500 (12$\frac{6}{7}$ days.
$$350$$

$$1000$$
$$700$$

$$\frac{300}{350} = \frac{6}{7}$$

184. The operations in questions of the Single Rule of Three may often be simplified by the process of *analysis*. Thus,

1. If 5 barrels of flour cost $35, what will 40 barrels cost?

OPERATION. Since 5 barrels cost $35, 1 barrel will cost $\frac{35}{5} = 7$ dollars; and 40 barrels will be worth $40 \times 7 = 280$ dollars.

The analysis consists in this example in finding the value of a single barrel of flour when that of 5 is known. This value is expressed by the ratio $\frac{35}{5}$, and thus ascertaining the value of 1 barrel, that of 40 is obtained by simple multiplication.

2. If a man travel 100 miles in 5 days, how many miles will he travel in 25 days at the same rate?

OPERATION.—If he travels 100 miles in 5 days, he will travel $\frac{100}{5} = 20$ miles in 1 day, and $25 \times 20 = 500$ miles in 25 days.

In this example the analysis shows that the man must travel 20 miles per day, and therefore he travels 500 miles in 25 days.

3. If 6 men consume 12 barrels of flour in 1 year, how much will 50 men consume in the same time?

OPERATION.—If 6 men consume 12 barrels in 1 year, 1 man will consume $\frac{12}{6} = 2$ barrels in the same time, and 50 men will require $50 \times 2 = 100$ barrels.

Here, by analysis, we find how much 1 man will consume at the given rate, and multiply the result by the number of men for the answer.

Q. How may questions in the single rule of three be simplified? How is the analysis applied in the first example? When you know the cost of a single thing, how may you find the cost of many? If 1 hat cost $3, what will 10 hats cost? 20? 30? 40? 50? How is the analysis applied in the second example? When you know the distance a man travels in 1 day, how may you find the distance travelled in 2, 3, 4, 5, &c. days? If he travel 10 miles a day, how far will he travel in 5 days? 10 days? 20? How is analysis applied in the third example? When you know the quantity of flour 1 man will consume in a given time, how may you find the quantity consumed by 2, 3, 4, &c. in the same time? If 1 man consume ½ barrel of flour in 3 mo. much will 10 men consume in the same time? 30 men?

EXAMPLES.

1. If 50 barrels of flour cost $250, what will 10 barrels cost ?   Ans. $50.

2. What will a barrel of sugar cost weighing 217 lbs. at 8¼ cts. per lb. ?   Ans. $18.44½.

3. What is the value of 3 barrels of molasses at 25 cts. per gallon ?   Ans. $23.62½.

4. If 80 barrels of flour will support 100 men 40 days, how long will they subsist 25 men ?   Ans. 160 days.

5. If 5 men can do 100 yards of work in 24 days, how many men will do the same work in 15 days ?
               Ans. 8 men.

6. A man's annual income is $1460 : what is it per day ?   Ans. $4.

7. How many yards of carpet that is 3 feet wide will cover a floor that is 27 feet long and 20 feet wide ?
               Ans. 60 yards.

8. A garrison of 536 men have provisions for 12 months : how long will these provisions last if the garrison be increased to 1124 men ?   Ans. 174$\frac{864}{1124}$ days.

9. A piece of work is done in 12 days by working 4 hours per day : what time will be required to finish it, by working 6 hours per day ?   Ans. 8 days.

10. How much corn can be bought for 80 dollars at 37½ cents per bushel ?   Ans. 213⅓ bushels.

11. A person fails owing $2000, but has in goods, money and debts due him $427.50 : how much will he pay his creditors per dollar, if he deliver these to them ?
               Ans. 21⅜ cts.

12. A person's income is $500 per year, and he spends daily $1.12½, how much will he save at the year's end ?
               Ans. $89.37½.

13. The governor of a besieged town, having provisions 54 days, at the rate of 1½ lbs. per day for each person,

is desirous to prolong the siege to 80 days, in the hope of succour: what must be the allowance for this purpose?

Ans. $1\frac{1}{80}$ lbs.

14. A man pays board at $4 per week: how long will $100 board him? Ans. 25 weeks.

15. The salary of the President of the United States is $25,000: how much may he spend per month to save $5000 at the year's end? Ans. 1666\frac{2}{3}$.

16. A soldier steps 2 yards in 3 steps: how many yards will he march in 160 steps? Ans. 106 yds. 2 ft.

CONTRACTIONS OF THE SINGLE RULE OF THREE.

185. When the price of a single thing is expressed in exact parts of a higher unit, the operations of the Rule of Three may be very much contracted by dividing by the number expressing the number of these parts, and the result will be the answer in terms of the higher unit.—Thus,

1. What is the value of 100 bushels of corn at 50 cents per bushel?

Since 50 cents are $\frac{1}{2}$ of a dollar, 100 bushels of corn will cost $\frac{1}{2}$ as much at 50 cents per bushel as at $1 per bushel. But 100 bushels at $1 per bushel cost $100: at 50 cents, they must therefore cost 1\frac{100}{2}$=$50.

OPERATION.

cts. $
$50=\frac{1}{2}$ 2) 100

50 cost in dolls.

2. What is the value of 100 bushels of wheat at 75 cents per bushel?

As 75 cts. do not form an exact part of $1, we first find how much 100 bushels of wheat will cost at 50 cts., and then what they will cost at 25 cts., and add the two results together for the value of the 100 bushels at 75 cts. Since 25 cts. are $\frac{1}{4}$ of $1, we divide 100 by 4, and we get $25 for the co

OPERATION.

cts. $
$50=\frac{1}{2}$ 2 | 100

50 cost at 50 cts.
$25=\frac{1}{4}$ 25 cost at 25 cts.

$75 cost at 75 cts.

wheat at 25 cts. Dividing the cost at 50 cts. by 2, would also give the cost at 25 cts., since 25 cts.$=\frac{1}{2}$ of 50 cts.

3. What is the value of 125 gallons of wine at 10*s*. 4*d*. per gallon?

OPERATION.

10 sh.$=\frac{1}{2}\pounds$　　　　2	125
4*d*.$=\frac{1}{30}$ of 10 sh. · 30	62 10 cost at 10*s*.
	2 1 8 " " 4*d*.

Ans. £64 11s. 8d. cost at 10s. 4d.

10 shillings being $\frac{1}{2}$ of £1, we first divide by 2, which gives the cost of the wine at 10 shillings; 4*d*. being $\frac{1}{3}$ of 1 shilling, is $\frac{1}{30}$ of 10 shillings; and dividing the cost at 10 shillings, by 30, we get the cost at 4*d*. Adding the two results, we get the answer.

Since 4*d*. is $\frac{1}{60}$ of £1, the same result would have been obtained by dividing 125 by 60, the cost being still £2 1*s*. 8*d*.

4. What is the value of 4 cwt. 3 qr. 14 lb. of sugar at $5.50 per cwt.?

Multiplying $5.50 by 4, we get the cost of 4 cwt. We then divide the cost of 1 cwt. by 2 for the cost of 2 qrs., since 2 qr. $=\frac{1}{2}$ cwt. Divide now the cost of 2 qr. by 2 for the cost of 1 qr., since 1 qr.$=\frac{1}{2}$ of 2 qr.; and this again by 2 for the cost of 14 lb., since 14 lb.$=\frac{1}{2}$ qr. The sum of these separate results is the answer.

OPERATION.

2 qr.$=\frac{1}{2}$ cwt.　　2	$5.50
	4
2 qr.$=\frac{1}{2}$ cwt.	22.00 cost of 4 cwt.
1 qr.$=\frac{1}{2}$ of 2 qr. 2	2.75 " 2 qr.
14 lb.$=\frac{1}{2}$ of 1 qr. 2	1.37$\frac{1}{2}$ " 1 qr.
	68$\frac{3}{4}$ " 14 lb.

Ans. 26.81\frac{1}{4}$, cost of 4 cwt. 3 qr. 14 lbs.

A number which is an exact part of another number is said to be an *aliquot* part of it: 50 cents, 25 cents, 10 cents, are aliquot parts of $1; since 50 cents is contained in $1, ? *times*; 25 cents, 4 times; 10 cents, 10 times, &c.

It is often convenient to have tables of aliquot parts of compound numbers. The following are most in use, and should be carefully committed to memory:—

TABLES OF ALIQUOT PARTS.

MONEYS.					
Cents.	$	Shillings.	£	Pence.	Shilling.
50 =	$\frac{1}{2}$	10 =	$\frac{1}{2}$	6 =	$\frac{1}{2}$
33$\frac{1}{3}$ =	$\frac{1}{3}$	6s. 8d. =	$\frac{1}{3}$	4 =	$\frac{1}{3}$
25 =	$\frac{1}{4}$	5 =	$\frac{1}{4}$	3 =	$\frac{1}{4}$
20 =	$\frac{1}{5}$	4 =	$\frac{1}{5}$	2 =	$\frac{1}{6}$
12$\frac{1}{2}$ =	$\frac{1}{8}$	3s. 4d. =	$\frac{1}{6}$	1$\frac{1}{2}$ =	$\frac{1}{8}$
6$\frac{1}{4}$ =	$\frac{1}{16}$	2s. 6d. =	$\frac{1}{8}$	1 =	$\frac{1}{12}$
5 =	$\frac{1}{20}$	1s. 8d. =	$\frac{1}{12}$		

WEIGHT.				TIME.			
Lbs.	Cwt.	Cwt.	Ton.	Month.	Year.	Days.	Month.
56 =	$\frac{1}{2}$	10 =	$\frac{1}{2}$	6 =	$\frac{1}{2}$	15 =	$\frac{1}{2}$
28 =	$\frac{1}{4}$	5 =	$\frac{1}{4}$	4 =	$\frac{1}{3}$	10 =	$\frac{1}{3}$
14 =	$\frac{1}{8}$	4 =	$\frac{1}{5}$	3 =	$\frac{1}{4}$	7$\frac{1}{2}$ =	$\frac{1}{4}$
7 =	$\frac{1}{16}$	2 =	$\frac{1}{10}$	2 =	$\frac{1}{6}$	6 =	$\frac{1}{5}$
						5 =	$\frac{1}{6}$
						3 =	$\frac{1}{10}$

Q. When the price of a single thing is expressed in exact parts of a higher unit, how may the operation of the rule of three be contracted? If the price of a single thing be 50 cents, by what would you divide? Why? What unit would the answer be? By what would you divide if the price were 25 cents? When is a number an aliquot part of another? Is 50 cts. an aliquot part of $1? What part? Is 25 cts.? Why? What part? Is 37½ cts. an aliquot part of $1? How would you apply this contraction in such a case? What number of cents are aliquot parts of $1? What part of $1 is each? What number of shillings is an aliquot part of £1? What part is each? Pence of shilling? What number of pounds is an aliquot part of 1 cwt.? What part is each? What number of cwts. is an aliquot part of 1 ton? What number is each? How many months form aliquot parts of 1 year? How many days form aliquot parts of 1 month?

EXAMPLES.

1. What is the value of 116½ yards of cloth at £2 2s. 6d. per yard?

By the table it will be seen 2*s.* 6*d.* = ⅛ of £1 : dividing £116, which will be the cost of 116 yds. at £1, by 8, we have £14 10*s.* as the cost at 2*s.* 6*d.*

OPERATION.

2*s.* 6*d.* = ⅛ of £1   8 | 116
                            | 2
                            | ————————
                            | 232 cost at £2.
                            | 14 10 " 2*s.*6*d.*
cost of ⅛ yd. | 1 01 3
                            —————————
                            £247 11 3.

2. What will 720¼ yards of cloth cost at 6*s.* 8*d.* per yard?   Ans. £240 1*s.* 8*d.*

3. What will 213 yards of cloth cost at £1 13*s.* 4½*d.* per yard?   Ans. £355 8*s.* 10¼*d.*

4. What will 112 cwt. 3 qrs. 14 lb. of sugar come to at $12.25 per cwt.?   Ans. $1382.71⅞.

5. What is the value of 10 boxes of soap, each box containing 30 lbs., at 6¼ cents per lb.?   Ans. $18.75.

6. What will be the cost of 275 yards of cloth at $1.37½ per yard?   Ans. $378.12¼.

7. What is the value of 416 gallons of molasses at 62½ cents per gallon?   Ans. $260.

8. What will 200 barrels of flour cost at $5.56¼ per barrel?   Ans. $1112.50.

9. What will 75 barrels of beef cost at $9.87½ per barrel?   Ans. $740.62½.

10. What will 100 tons of iron cost at $3.93¾ cents per cwt.?   Ans. $7875.

11. What will 16 cwt. 2 qr. 18 lb. of sugar cost at $7.75 per cwt.?   Ans. 129.12_{\tau}\frac{6}{3}$.

12. What will 22,000 feet of lumber cost at $9.25 per thousand feet?   Ans. $203.50.

COMPOUND RULE OF THREE.

186. *The* COMPOUND RULE OF THREE *embraces a class of questions which requires two or more statements.*

1. Thus, if 9 men can build 36 feet of wall in 4 days, how many feet of wall will 19 men build in 11 days?

In this example, we first ascertain how many feet of wall 19 men can build while 9 men are building 36 feet. It is plain, 9 men will be to 19 men, as 36, the number of feet built by 9 men, is to 76, the number of feet built by 19 men.

1st OPERATION.

men. men. feet. feet.
9 : 19 :: 36 : Ans.
19
———
324
36
———
9) 684
———
76 feet.

We now ascertain how many feet 19 men can build in 11 days, when it is known they can build 76 feet in 4 days.

Since 19 men can build 76 feet of wall in 4 days, it is evident 4 days will be to 11 days, as 76, the number of feet built in 4 days, is to 209, the number of feet built in 11 days: 209 feet is therefore the required answer.

2d OPERATION.

days. days. feet. feet.
4 : 11 :: 76 : Ans.
11
———
4) 836
———
Ans. = 209 feet.

2. If 5 persons consume 10 barrels of flour in 1 year, how many barrels will 20 persons consume in 7 years at the same rate?

We first ascertain how many barrels 20 persons will use in 1 year at the given rate. Here, 5 persons will be to 20, as 10, the number of barrels used by 5 persons, is to 40, the number required.

1st OPERATION.

persons. persons. bar. bar.
5 : 20 :: 10 : Ans.
20
———
5) 200
———
40 bbls.

Now, since 20 persons consume 40 barrels in 1 year, by a simple proportion we can find how many they will consume in 7 years. They will require 280 barrels.

2d OPERATION.

year. year. barrels. barrels.
$$1 : 7 :: 40 : Ans.$$
$$7$$

Ans. $= 280$ barrels.

3. If 12 men can dig 100 feet of ditch in 5 days, how many men will be required to dig 200 feet in 8 days?

Here, we first find the number of days 12 men will require to dig 200 feet. It is plain, 100 feet will be to 200 feet, as 5, the number of days required to dig 100 feet, is to 10, the number of days required to dig 200 feet.

1st OPERATION.

feet. feet. days. days.
$$100 : 200 :: 5 : Ans.$$
$$200$$

$$100) 1000$$
---
10 days.

It is now required to find the number of men necessary to dig a ditch in 8 days, which 12 men are 10 days digging.

Since the work has to be done in a shorter time, more men will be required. The fourth term of the proportion must then be greater than the third term, which is 12, and 10 days will be the second term. The answer is 15 men.

2d OPERATION.

days. days. men. men.
$$8 : 10 :: 12 : Ans.$$
$$10$$

$$8) 120$$

Ans. $= 15$ men.

Q. What kind of questions does the compound rule of three embrace?

187. The questions which we have just examined, as well as all others of the same kind, might be solved by another mode.

Thus, in the first example, if 9 men can build 36 feet of wall in 4 days, they will build $\frac{36}{4} = 9$ feet in 1 day, and $9 \times 11 = 99$ feet in 11 days. The question then reduces itself to find how many feet 19 men can build while 9 men are building 99 feet.

OPERATION.

men. men.   feet.   feet.
9 : 19 :: 99 : Ans.
                19
            ———
            891
             99
            ———
          9) 1881
            ———

By a simple proportion we find that 9 men is to 19, as 99, the number of feet built by 9 men, is to 209, the number of feet built by 19 men. The same result that we found before.

Ans. = 209 feet.

In the second example, if 5 persons consume 10 barrels of flour in 1 year, 1 person will use $\frac{10}{5} = 2$ barrels in 1 year, and $2 \times 7 = 14$ barrels in 7 years. If 1 person consume 14 barrels of flour in 7 years, how many will 20 persons consume in the same time?

OPERATION.

person. person.   barrel.   barrel.
1 : 20 :: 14 : Ans.
                20

We find by a simple proportion that 20 persons will consume 280 barrels, as found before.

Ans. = 280 bbls.

In the third example, if 12 men dig 100 feet in 5 days, they will dig $\frac{100}{5} = 20$ feet in 1 day. But the required number of men have to dig 200 feet in 8 days; that is, they must dig $\frac{200}{8} = 25$ feet per day.

OPERATION.

feet.   feet.   men.   men.
20 : 25 :: 12 : Ans.
             25
           ———
       20) 300
           ———
       Ans. = 15 men.

We have then the proportion 20, the number of feet 12 men will dig in 1 day, to 25, the number of feet dug by the required number of men in the same time, as 12 men is to the required number of men, which we find to be 15, as before found.

The process by which these solutions are made is called *analysis,* and consists in discovering from the nature of the question the relation which a single thing bears to its corresponding term. This operation can often be made mentally.

Q. What is the name given to the process which has just been explained for working questions in the compound rule of three? In what does it consist? How do you illustrate this by the first example? cond example? Third example?

EXAMPLES.

1. If a person travel 300 miles in 10 days of 12 hours each, in how many days of 16 hours each will he travel 600 miles?   Ans. 15 days.

2. If a family of 9 persons spend $480 in 8 months, how much will serve a family of 24 persons living at the same rate for 16 months?   Ans. $2560.

3. If 28 dollars be the wages of 4 men for 7 days, what will be the wages of 14 men for 10 days?
                              Ans. $140.

4. If 120 bushels of corn can serve 14 horses 56 days, how many days will 94 bushels serve 6 horses?
                              Ans. $102\frac{14}{18}$ days.

5. If 3000 lb. of beef serve 340 men 15 days, how many pounds will serve 120 men for 25 days?
                              Ans. 1764 lb. $11\frac{14}{17}$ oz.

6. If a barrel of potatoes last a family of 8 persons 12 days, how many barrels will be required for 16 persons for 1 year?   Ans. $60\frac{5}{6}$ barrels.

7. If 180 men in 6 days of 10 hours each, can dig a trench 200 yards long, 3 wide, and 2 deep, in how many days, 8 hours long, will 100 men dig a trench 360 yards long, 4 wide, 3 deep?   Ans. 15 days.

APPLICATION OF THE PRECEDING RULES.

Find the total value of the principal annual products of the several States, as determined by the census of 1840, as follows :—

MAINE.

6,122 tons of cast iron at $3.25 per cwt.	$397,930.00
16,666⅔ sacks of salt, (each sack containing 3 bushels,) at 35 cents per bushel,	17,500.00
848,166 bushels of wheat, at 75 cents per bushel,	636,124.50
1,076,409 bushels of oats, at 20¼ cents per bushel,	220,863.84¼

1,465,551 lbs. of wool, at 37½ cts. per lb.	549,581.62½
3,464,073 barrels of potatoes of 3 bushels each, at 12½ cts. per bushel,	1,299,027.37½
691,358 tons of hay, at $1.10 per cwt.,	15,209,876.00
279,156 quintals of fish of 100 lbs. each, at 2¼ cts. per lb.,	628,101.00
180,868,300 feet of lumber, at $1.06 per hundred feet,	1,917,203.98
1869 hhds. of sperm oil, at 70 cts. per gallon,	82,422.90
7129 boxes tallow candles of 30 lbs. each, at 12½ cts. per lb.,	26,733.75
	$20,985,164.97½

NEW HAMPSHIRE.

29,920 bushels of bituminous coal, at $1.15 per chaldron,	$
1,296,114 bushels of oats, at ¼ of a dollar per bushel,	$
105,103 bushels of buckwheat, at $1½ per barrel of 3 bushels each,	$
496,107 tons of hay, at $1⅛ per cwt.,	$
1,666 boxes of sperm candles, of 30½ lbs. each, at 31¼ cts per lb..	$
Ans.	$11,555,821.01 +

MASSACHUSETTS.

9,332 tons of cast iron, at $3¼ per cwt.,	$
6,004 tons of bar iron, at 4½ cts. per lb.,	$
125,532 barrels of salt, of 3 bushels each, at 40 cts. per bushel,	$
389,715 quintals of fish of 100 lbs., at 2½ cts. per lb.,	$
124,755 barrels of pickled fish, of 200 lbs. each, at 2½ cts. per lb.,	$
251,208 boxes of soap, of 50 lbs. each, at 3¼ cts. per lb.,	
41,915 boxes of tallow candles, of 30 lbs. each, at 10½ cts. per lb.	$

72,090 boxes of sperm candles, of 30 lbs.
   each, at 32¾ cts. per lb., $

569,395 tons of hay, at 75 cts. per cwt., $

57,634 hogsheads of sperm oil, at 65 cts.
   per gallon, $

53,408 hogsheads of fish oil, at 31¼ cts.
   per gallon, $
                                      $16,218,106.

RHODE ISLAND.

4,126 tons of cast iron, at $1¼ per quarter, $

911,973 bushels of potatoes, 37¼ cts. per
   barrel of 3 bushels each, $

63,449 tons of hay, at 87½ cts. per cwt., $

24,741 boxes of soap, of 50 lbs. each, at
   3¾ cts. per lb., $
                                      $1,687,093.5(

CONNECTICUT.

3,414,238 bushels of potatoes, at 50 cts.
   per hamper of 2½ bushels each, $

30,302 hhds. of fish oil, at 37½ cts. per
   gallon, $

426,704 tons of hay, at 87¼ cts. per cwt., $

737,424 bushels of rye, at 81¼ cts. per
   bushel, $

14,693 boxes of tallow candles, of 30 lbs.
   each, at 10¾ cts. per lb., $
                                      $9,143,882.2'

VERMONT.

36,992 bales of wool, of 100 lbs. each, at
   42 cts. per lb., $

8,869,751 bushels of potatoes, at 62½ cts.
   per hamper of 2½ bushels each, $

836,739 tons of hay, at 68¾ cts. per cwt., $

1,681,819 sheep, at $4½ per pair, $
                                      $19,060,355.'

New York.

29,088 tons of cast iron, at 3¾ cts. per lb., $

53,693 tons of bar iron, at $4¾ per cwt., $

9,444 pigs of lead, of 70 lbs. each, at $3¼
   per 100 lbs., $

955,961 barrels of salt, of 3 bushels each,
   at 40 cts. per bushel, $

12,286,418 bushels of wheat, at $1.06¼
   per bushel, $

20,675,847 bushels of oats, at 75 cts. per
   bag of 3 bushels each, $

762,628 barrels of buckwheat, of 196 lbs.
   each, at $1½ per 100 lbs., $

2,194,457 barrels of corn, of 5 bushels
   each, at 55 cts. per bushel, $

30,123,614 bushels of potatoes, at 75 cts.
   the hamper of 2½ bushels each, $

3,127,047 tons of hay, at $1¼ per cwt., $

40,302 barrels of fish oil, at 43¾ cts. per
   gallon, $

389,130,200 feet of lumber, at $12¼ per
   1000 feet, $

170,569 boxes of soap, of 70 lbs. each, at
   $3¼ per 100 lbs., $

134,326 boxes of tallow candles, of 30 lbs.
   each, at 12½ cts. per lb., $

98,453 bales of wool, of 100 lbs. each, at
   50 cts. per lb., $

$133,691,947.13¼

New Jersey.

11,114 tons of cast iron, at $1¼ per quar-
   ter, $

7,171 tons of bar iron, at $5¾ per cwt., $

334,861 tons of hay, at 37½ cts. per quarter, $

2,539 barrels of fish oil, at 43¾ cts. per
   gallon, $

H

6,903 boxes of soap, of 70 lbs. each, at $3⅜ per 100 lbs., $
12,418 boxes of tallow candles, of 30 lbs. each, at $2¼ per quarter, $

$12,009,347.6

PENNSYLVANIA.

98,395 tons of cast iron, at $3¼ per cwt., $
87,244 tons of bar iron, at $1 per quarter, $
859,686 tons of anthracite coal, of 28 bushels each, at 20 cts. per bushel, $
11,620,654 bushels of bituminous coal, at $6¼ per chaldron, $
549,478 bushels of salt, at $2¼ per sack of 3 bushels, $
13,213,077 bushels of oats, at 37½ cts. per bushel, $
20,641,819 bushels of wheat, at $1.12½ per bushel, $
1,311,643 tons of hay, at $1.10 per cwt., $
72,824 boxes of soap, of 70 lbs. each, at $4¾ per cwt., $
77,228 boxes of tallow candles, of 30 lbs. each, at $14½ per cwt., $

$78,167,129.9

DELAWARE.

17 tons of cast iron, at $3¼ per cwt., $
449 tons of bar iron, at $3¾ per cwt., $
789 hhds. of sperm oil, at 76½ cts. per gallon, $
2263 hhds. of fish oil, at 31¼ cts. per gallon, $
22,483 tons of hay, at $1.06¼ per cwt., $

$595,122.41+

MARYLAND.

8,876 tons of cast iron, at $4¾ per cwt., $
7,900 tons of bar iron, at 87½ cts. per *quarter*, $

222,000 bushels bituminous coal, at $7¾
   per chaldron, $
8,233,086 bushels of corn, at $4¼ per
   barrel of 5 bushels, $
106,687½ tons of hay, at $1.10 per cwt., $
24,816 hhds. of tobacco, of 1000 lbs. each,
   at $5¾ per 100 lbs., $

$12,216,179.76⅔

VIRGINIA.

18,810¼ tons of cast iron, at $4½ per cwt., $
5,886 tons of bar iron, at 5¼ cts. per lb., $
12,552 pigs of lead, of 70 lbs. each, at $7
   per cwt. $
10,622,345 bushels of bituminous coal, at
   $7¾ per chaldron, $
581,872 barrels of salt, of 3 bushels each,
   at 37½ cts. per bushel, $
10,109,716 bushels of wheat, at $1.05 per
   bushel, $
34,577,591 bushels of corn, at $4½ per
   barrel of 5 bushels, $
364,708½ tons of hay, at 75 cts. per 100
   lbs., $
25,594¼ tons of hemp and flax, at $7½ per
   cwt., $
75,347 hhds. of tobacco, of 1000 lbs. each,
   at $6¼ per 100 lbs., $
11,648 bales of cotton, of 300 lbs. each, at
   6¼ cts. per lb., $

$62,198,643.42+

NORTH CAROLINA.

23,893,763 bushels of corn, at $3¾ per
   barrel of 5 bushels, $
101,369 tons of hay, at 75 cts. per 100 lbs., $
26,772 hhds. of tobacco, of 1000 lbs. each,
   at $5½ per 100 lbs., $
28,203 tierces of rice, of 100 lbs. each, at
   $3½ per 100 lbs., $

173,087 bales of cotton, of 300 lbs. each, at 7½ cts. per lb.,　　　$

73,350 barrels of pickled fish, at $3¼ per barrel,　　　$

593,451 barrels of tar, pitch, &c., at $4¾ per barrel,　　　$

50,676,600 feet of lumber, at $1½ per 100 feet,　　　$

　　　　　　　　　　　　　$28,159,290.20

SOUTH CAROLINA.

1,250 tons cast iron, at 3¼ cts. per lb.,　　$

1,165 tons bar iron, at $5.90 per cwt.,　　$

14,722,805 bushels of corn, at $2.50 per barrel of 5 bushels,　　　$

605,908 tierces of rice, of 100 lbs. each, at $3.25 per 100 lbs.,　　　$

205,700 bales of cotton, of 300 lbs. each, at 7½ cts. per lb.,　　　$

53,768,400 feet of lumber, at $15 per 1000 feet,　　　$

　　　　　　　　　　　　　$14,993,849.50

GEORGIA.

20,905,122 bushels of corn, at $3¾ per barrel of 5 bushels,　　　$

163 hhds. of tobacco, of 1000 lbs. each, at $7¼ per 100 lbs.,　　　$

123,847 tierces of rice, of 100 lbs. each, at $4½ per 100 lbs.,　　　$

544,641 bales of cotton, of 300 lbs. each, at 6¼ cts. per lb.,　　　$

　　　　　　　　　　　　　$26,459,989.25

ALABAMA.

20,947,004 bushels of corn, at $3.30 per barrel of 5 bushels,　　　$

273 hhds. of tobacco, of 1000 lbs. each, at 7 cts. per lb.,　　　$

1,490 tierces of rice, of 100 lbs. each, at
   5 cts. per lb., $

390,463 bales of cotton, of 300 lbs. each,
   at 5½ cts. per lb., $

                                     $20,294,222.14

MISSISSIPPI.

13,161,237 bushels of corn, at $4.50 per
   barrel of 5 bushels, $

7,772 tierces of rice, of 100 lbs. each, at
   3½ cts. per lb., $

644,672 bales of cotton, of 300 lbs. each,
   at $7¼ per 100 lbs., $

                                     $25,893,931.30

LOUISIANA.

1,400 tons of cast iron, at $3½ per cwt., $

1,366 tons of bar iron, at $4¼ per cwt., $

24,651 tons of hay, at 75 cts. per 100 lbs., $

120 hhds. of tobacco, of 1000 lbs. each,
   at $8¼ per 100 lbs., $

36,045 tierces of rice, of 100 lbs. each, at
   3½ cts. per lb., $

508,517 bales of cotton, of 300 lbs. each,
   at 7⅛ cts. per lb., $

119,947 hhds. of sugar, of 1000 lbs. each,
   at $4¼ per 100 lbs., $

31,460 boxes of soap, of 70 lbs. each, at
   5¼ cts. per lb., $

116,667 boxes of tallow candles, of 30 lbs.
   each, at 14¼ cts. per lb., $

                                 $17,266,962.66½

TENNESSEE.

16,128¼ tons of cast iron, at 3¼ cts. per
   lb., $

9,673 tons of bar iron, at 4¼ cts. per lb., $

44,986,188 bushels of corn, at $2½ per
   barrel of 5 bushels, $

31,233 tons of hay, at 62½ cts. per cwt., $

29,550 hhds. of tobacco, of 1000 lbs. each, at $5¼ per 100 lbs., $

92,337 bales of cotton, 300 lbs. each, at 5¼ cts. per lb., $

2,926,607 hogs, of 200 lbs. each, at $2¼ per 100 lbs., $

                                                    $41,318,139.75

KENTUCKY.

29,206 tons of cast iron, at $3.50 per cwt., $

3,637 tons of bar iron, at $4.75 per cwt., $

588,167 bushels of bituminous coal, at $6¼ per chaldron, $

1,008,240 sheep, of 30 lbs. each, at 3½ cts. per lb., $

2,310,533 hogs, of 200 lbs. each, at $2¼ per 100 lbs., $

39,847,120 bushels of corn, at $2.75 per barrel of 5 bushels, $

88,306 tons of hay, at 87½ cts. per 100 lbs., $

53,436 hhds. of tobacco, of 1000 lbs. each, at $7½ per 100 lbs., $

32,606 boxes of soap, of 70 lbs. each, at 4½ cts. per lb., $

219,695 bushels of salt, at $2¼ per sack of 3 bushels, $

9,992¼ tons of hemp and flax, at $6.75 per cwt., $

                                                    $43,218,945.31

OHIO.

35,236 tons of cast iron, at $3.20 per 100 lbs., $

7,466 tons of bar iron, at 4¼ cts. per lb., $

3,513,409 bushels of bituminous coal, at $7 per chaldron, $

297,350 bushels of salt, at $2½ per sack of 3 bushels, $

16,571,661 bushels of wheat, at $1.04½ per *bushel,* $

33,668,144 bushels of corn, at $2.75 per barrel of 5 bushels, $

1,022,037 tons of hay, at $1.06¼ per cwt., $

9,080¼ tons of hemp and flax, at 6¼ cts. per lb., $

5,942 hhds. of tobacco, of 1000 lbs. each, at $7¼ per 100 lbs., $

51,472 boxes of soap, of 70 lbs. each, at 3¼ cts. per lb., $

77,282 boxes of tallow candles, of 30 lbs. each, at 10½ cts. per lb., $

$63,792,160.10 +

INDIANA.

810 tons of cast iron, at $3¼ per cwt., $

242,040 bushels bituminous coal, at $6.25 per chaldron, $

28,155,887 bushels of corn, at $2.44 per barrel of 5 bushels, $

178,029 tons of hay, at 75 cts. per cwt., $

8,605½ tons of hemp and flax, at $5½ per cwt., $

1,820 hhds. of tobacco, of 1000 lbs. each, at $5.30 per 100 lbs., $

$17,548,243.42 +

ILLINOIS.

125,071 pigs of lead, of 70 lbs. each, at 3¼ cts. per lb., $

132 tons of anthracite coal, 28 bushels to a ton, at 20 cts. per bushel, $

424,187 bushels bituminous coal, at $4.75 per chaldron, $

22,634,211 bushels of corn, at $2¼ per barrel of 5 bushels, $

164,932 tons of hay, at 25 cts. a quarter, $

560 hhds. of tobacco, of 1000 lbs. each, at $5.60 per cwt., $

669 bales of cotton, of 300 lbs. each, at 7½ cts. per lb., $

$13,890,

MISSOURI.

75,649 pigs of lead, of 70 lbs. each, at
$4¼ per cwt., $

249,302 bushels of bituminous coal, at $7½
per chaldron, $

17,332,524 bushels of corn, at $2.75 per
barrel of 5 bushels, $

18,010¾ tons of hemp and flax, at $5½ per
cwt., $

9,068 hhds. of tobacco, of 1000 lbs. each,
at $7¼ per cwt., $

_____

$12,365,471.02·

ARKANSAS.

5,500 bushels of bituminous coal, at $6¾
per chaldron, $

4,846,632 bushels of corn, at $2¾ per bar-
rel of 5 bushels, $

148 hhds. of tobacco, of 1000 lbs. each, at
5½ cts. per lb., $

54 tierces of rice, of 100 lbs. each, at 3½
cts. per lb., $

20,095 bales of cotton, of 300 lbs. each,
at 7¼ cts. per lb., $

_____

$3,112,074.10

MICHIGAN.

601 tons of cast iron, at 3¼ cts. per lb., $

2,157,108 bushels of wheat, at 87¼ cts.
per bushel, $

464,257 barrels of corn, of 5 bushels each,
at 53½ cts. per bushel, $

180,805 tons of hay, at $1.10½ per cwt., $

_____

$6,063,900.27½

FLORIDA TERRITORY.

75 hhds. of tobacco, of 1000 lbs. each, at
$8½ per cwt., $
1,600 tierces of rice, of 300 lbs. each, at
3¼ cts. per lb., $
40,368 bales of cotton, of 300 lbs. each,
at 7¾ cts. per lb., $
_____
$959,847.75

WISCONSIN TERRITORY.

216,133 pigs of lead, of 70 lbs. each, at
$3¾ per 100 lbs., $567,349.12½

IOWA TERRITORY.

7,157 pigs of lead, of 70 lbs. each, at 3½
cts. per lb., $17,534.65

DISTRICT OF COLUMBIA.

55 hhds. of tobacco, of 1000 lbs. each, at
5¾ cts. per lb., $3,162.50

Q. What are the principal productions of Maine? New Hampshire? Massachusetts? Vermont? Connecticut? Rhode Island? New York? New Jersey? Pennsylvania? Delaware? Maryland? Virginia? &c. &c. Which are the largest corn-growing States? Wheat? What States produce tobacco? Cotton? Rice? Where is lead found? Iron? What State produces sugar-cane? Coal? Where is bituminous coal found? Anthracite? What are the principal productions of your State?

2. By an act of Congress of 1841, the nett proceeds of the public lands sold in the half year ending 30th of June, 1842, amounting to $562,144.18, were to be distributed among the several States according to their federal population: what was the share of each State and Territory, according to the following federal numbers?

H 2

States.	Federal Population.	Share.
1. Maine, ...	501,793	Ans. $17,554.90
2. New Hampshire,.	284,574	Ans. 9,955.64
3. Massachusetts, ...	737,699	Ans. 25,807.92
4. Rhode Island, ...	108,828	Ans. 3,807.28
5. Connecticut, ...	310,008	Ans. 10,845.43
6. Vermont, ...	291,948	Ans. 10,213.61
7. New York, ...	2,428,919	Ans. 84,974.15
8. New Jersey, ...	373,036	Ans. 13,050.42
9. Pennsylvania, ...	1,724,007	Ans.
10. Delaware, ...	77,043	Ans.
11. Maryland, ...	434,124	Ans.
12. Virginia, ...	1,060,202	Ans.
13. North Carolina,..	655,092	Ans.
14. South Carolina,..	463,583	Ans.
15. Georgia, ...	579,014	Ans.
16. Alabama, ...	489,343	Ans.
17. Mississippi, ...	297,567	Ans.
18. Louisiana, ...	285,030	Ans.
19. Tennessee, ...	755,986	Ans.
20. Kentucky, ...	706,925	Ans.
21. Ohio, ...	1,519,466	Ans.
22. Indiana, ...	685,865	Ans.
23. Illinois, ...	476,051	Ans.
24. Missouri, ...	360,406	Ans.
25. Arkansas, ...	89,600	Ans.
26. Michigan, ...	212,267	Ans.
27. Wisconsin, ...	30,941	Ans.
28. Iowa, ...	43,106	Ans.
29. Florida, ...	44,190	Ans.
30. Dist. of Columbia,	41,830	Ans.

| 16,068,447 | $562,144.18 |

2. The following table exhibits the annual products of industry in the United States by the census of 1840: find the proportional amount to each individual in each State in whole numbers, assuming the population as stated in *Art. 9.*

VALUE OF PRODUCTS FROM

States.	Agriculture.	Manufactures.	Commerce.	Mining.	Forest.	Fisheries.	
Maine,	$15,856,270	$5,615,303	$1,505,360	$227,376	$1,877,663	$1,280,713	Ans. 52
New Hampshire,	11,377,752	6,545,811	1,001,533	88,373	449,861	92,311	Ans. 68
Vermont,	17,879,155	6,665,425	758,869	369,468	430,224		Ans. 85
Massachusetts,	16,065,697	43,516,057	7,004,691	2,020,572	377,354	6,483,995	Ans. 103
Rhode Island,	2,198,309	8,640,625	1,294,956	162,410	44,510	652,312	Ans. 110
Connecticut,	11,971,776	12,778,963	1,988,281	890,419	181,575	907,723	Ans. 90
New York,	108,275,281	47,454,514	24,311,715	7,406,070	5,040,781	1,316,072	Ans. 79
New Jersey,	16,209,853	10,496,257	1,906,929	1,073,521	861,326	124,140	Ans. 79
Pennsylvania,	68,180,924	33,354,279	10,508,368	17,866,146	1,203,578	35,360	Ans. 76
Delaware,	3,196,440	1,536,879	986,257	54,555	13,119	181,285	Ans. 67
Maryland,	17,586,720	6,212,877	3,490,087	1,056,210	241,194	235,773	Ans. 61
Dist. of Columbia,	176,942	904,526	802,735			97,400	Ans. 45
Virginia,	59,085,821	8,349,218	5,299,451	3,321,629	617,760	95,173	Ans. 62
North Carolina,	26,975,831	2,053,697	1,922,284	372,486	1,446,108	251,792	Ans. 44
South Carolina,	21,558,691	2,248,915	2,632,421	187,608	549,626	1,275	Ans. 45
Georgia,	31,468,271	1,953,950	2,348,488	191,631	117,43	684	Ans. 52
Florida,	1,834,237	434,544	464,637	2,700	27,350	213,219	Ans. 54
Alabama,	24,696,513	1,732,770	2,273,267	81,310	177,465		Ans. 49
Mississippi,	26,494,565	1,565,790	1,453,686		205,297		Ans. 79
Louisiana,	22,351,375	4,087,655	7,996,996	165,280	71,751		Ans. 99
Arkansas,	5,096,757	1,145,309	420,635	18,595	217,469		Ans. 70
Tennessee,	31,660,180	2,477,193	2,239,478	1,371,381	235,179		Ans. 45
Mobile,	10,484,263	2,360,708	2,349,245	187,669	448,559		Ans. 41
Kentucky,	29,326,545	6,092,368	2,560,575	1,539,919	184,799		Ans. 49
Ohio,	37,802,001	14,588,091	5,050,316	2,442,682	1,013,063	10,595	Ans. 42
Indiana,	17,247,743	3,676,705	1,866,155	660,836	80,000	1,192	Ans. 34
Illinois,	13,701,406	3,243,981	1,493,425	299,272	249,841		Ans. 39
Michigan,	4,502,899	1,376,249	632,322	56,790	467,540		Ans. 33
Wisconsin,	568,105	204,692	189,957	384,603	430,580	27,663	Ans. 47
Iowa,	769,295	178,087	136,595	13,250	89,949		Ans. 27

Note.—The population of the District of Columbia in 1840 was 43,712; of Florida, 54,477; Wisconsin, 30,945; and Iowa, 43,112.

Q. Which State produces the most in agriculture? In manufactures? From the forest? Fisheries? Mining? Commerce? Which State produces least from agriculture? Manufactures? Commerce? Mining? Forest? Fisheries? In which State is the total proportional production to each individual the greatest? Which the least? How is it in your State? Name the principal agricultural States. The principal manufacturing States. Commercial States. Mining, &c. &c.

REDUCTION OF CURRENCIES.

188. FROM the want of a uniform system of currency of equal value in all parts of the world, it is necessary to explain the methods by which the reduction of currencies is effected between this and foreign countries.

REDUCTION OF CURRENCIES consists of two kinds, viz.

1st. *To reduce a sum in the currency of a foreign country to United States or Federal Currency.*

2d. *To reduce a sum in Federal currency to the currency of any foreign country.*

Q. Do all countries have the same currency? What reduction is necessary in consequence? How many kinds of reduction of currencies are there? What are they?

CASE I.

189. *To reduce a sum in the currency of a foreign country to United States or Federal Currency.*

Let us take for example the English currency.

The value of foreign coin is regulated by Congress. By a law of 1832, the English pound was fixed at $4.80.

Hence, $1 = \frac{10}{48}$ of £1 = $\frac{240}{48}$ pence = 50 pence.

We have therefore

$$
\begin{array}{cccc}
\pounds & \$ & \text{pence.} & \text{pence.} \\
1 & : 1 :: & 240 & : 50
\end{array}
$$

Hence £1 = $\frac{240}{50}$ = $\frac{24}{5}$.

£2 = $2 × $\frac{24}{5}$.

£3 = $3 × $\frac{24}{5}$, &c.

Hence, knowing the value of £1 to be $\frac{24}{5}$, we may reduce any number of pounds to dollars and cents, &c., by multiplying the given pounds by 24 and dividing by 5.

Thus, to reduce £175 to federal currency,

It is plain, that as £1 = $4.80, the same result would be obtained by multiplying the given pounds by $4.80. Thus, £175 × 4.80 = $840.00.

OPERATION.

$$
\begin{array}{r}
\pounds \\
175 \\
24 \\
\hline
700 \\
350 \\
\hline
5) 4200 \\
\hline
840 \text{ dollars.}
\end{array}
$$

Q. How is the value of foreign coin regulated ? What is the value of the English pound ? What fraction of $1 is £1 ? How may pounds be reduced to dollars and cents ? In what other way may the reduction be effected ?

190. Should the given amount be expressed in pounds, shillings and pence, the shillings and pence are brought to the decimal of a pound, and we then multiply as before. *Thus,*

Reduce £14 4s. 6d. to federal currency.

$$4s.\ 6d.=54d.=\tfrac{54}{240}\text{ of a }£=£.225$$

We first bring 4s. 6d. to the decimal of a pound, and then multiply the whole £14.225 by 24, and divide by 5. The answer is $68.28.8+.

$$£14\ 4s.\ 6d.=£14.225$$
$$24$$
$$\overline{}$$
$$56900$$
$$28450$$
$$\overline{}$$
$$5\ )\ 341400$$
$$\overline{}$$
Ans. $68.28.0

Q. If the sum contains pounds, shillings and pence, how do you proceed ?

191. In like manner, when we know the value of the standard foreign unit, we may readily find what part of $1 this unit is, and then reducing the given sum to the decimal of this unit, we multiply the result by the known fraction of $1, and we shall have the value of the given sum in dollars, cents, &c.

In the Canada currency, $1=60d. Hence,

$$£1 : \$1 :: 240d. : 60d.$$
$$£1=\$\tfrac{240}{60}=\$1\times 4.$$

Hence Canada money is reduced to federal money by multiplying the given pounds and decimals of a pound by 4.

Thus, reduce £30 10s. 4d. Canada currency to federal money.

Bringing the shillings and pence to the decimal of a pound, we multiply the result by 4, and the answer is $122.06.4+.

$$10s.\ 4d.=£.516+$$
$$£30\ 10s.\ 4d.=£30.516$$
$$4$$
$$\overline{}$$
Ans. $=$122.06.4+.$

Q. How may any foreign currency be brought to federal money ? What is the value of $1 in Canada currency ? What fraction of a dollar is £1 in this currency ? How is Canada currency brought to federal money ?

192. Before a uniform currency was established in the United States, each State had its own currency. These currencies are rapidly passing into disuse, but it may be proper to notice how the dollar was reckoned in each.

New York, Ohio,	$8s.=96d.=\$1$; hence $£1=\$\frac{240}{96}=\$2\frac{1}{2}$.
N. England States, Virginia, Kentucky, Tennessee,	$6s.=72d.=\$1$; hence $£1=\$\frac{240}{72}=\$3\frac{1}{3}$.
New Jersey, Pennsylvania, Delaware, Maryland,	$7s.\ 6d.=90d.=\$1$; hence $£1=\$\frac{240}{90}=\$2\frac{2}{3}$.
South Carolina, Georgia,	$4s.\ 8d.=56d.=\$1$; hence $£1=\$\frac{240}{56}=\$4\frac{2}{7}$.
North Carolina,	$10s.=120d.=\$1$; hence $£1=\$\frac{240}{120}=\$2$.

To reduce any of the above currencies to federal money, we have only to multiply the pounds and the decimal of a pound, by the value of £1 in the fraction of $1. Thus, New York currency is brought to federal money, by multiplying by $\frac{5}{2}$; Virginia, by multiplying by $\frac{10}{3}$; Pennsylvania, by $\frac{8}{3}$; South Carolina, by $\frac{30}{7}$; and North Carolina by 2.

Q. Did the States have originally the same currency? What was the value of $1 in New York? In Ohio? In Pennsylvania? Maryland? Delaware? New Jersey? The New England States? Virginia? Kentucky? Tennessee? Georgia? South Carolina? North Carolina? What fraction of $1 was £1 in each of these States? How may you reduce the currency of each of these States to federal money? Are these currencies much used?

193. The following table exhibits the value of foreign coin as fixed by Congress:

English pound, ...	$4.80
Livre of France, ...	$0.18½
Franc of France, ...	$0.18¾
Silver rouble of Russia, ...	$0.75
Florin or guilder of United Netherlands,	$0.40
Marc banco of Hamburg, ...	$0.33⅓
Real plate of Spain, ...	$0.10

Real vellon of Spain, ... $0.05
Milrea of Portugal, ... $1.24
Tale of China, ... $1.48
Pagoda of India, ... $1.84
Rupee of Bengal, ... $0.50

EXAMPLES.

1. Reduce £100 14s. 4d. English currency to federal money. Ans. $483.43.6+.

2. Reduce £1754 3s. 2½d. English currency, to federal money. Ans. $8419.96.8+.

3. Reduce £200 12s. 9d. English currency, to federal money. Ans. $963.05.7+.

4. Reduce £200 12s. 9d. Canada currency, to federal money. Ans. $802.54.8+.

5. Reduce £200 12s. 9d. New England currency, to federal money. Ans. $668.79+.

6. Reduce £200 12s. 9d. New York currency, to federal money. Ans. $501.59.2+.

7. Reduce £200 12s. 9d. Pennsylvania currency, to federal money. Ans. $535.03.2+.

8. Reduce £200 12s. 9d. Georgia currency, to federal money. Ans. $859.87.2+.

9. Reduce £200 12s. 9d. North Carolina currency, to federal money. Ans. $401.27.4.

CASE II.

194. *To reduce a sum in United States or federal money to the currency of any other country.*

For example, to reduce federal money to English currency —

£1 being equal to $4.80, $1 = $\frac{240}{48}$ pence = 50 pence.

Hence, £1 : $1 :: 240d. : 50d.

and $1 = £\frac{50}{240} = £\frac{5}{24}$.

$2 = £2 × \frac{5}{24}$.

$3 = £3 × \frac{5}{24}$.

*Hence, to convert any currency to federal money, find
the value of the dollar in terms of the standard unit of
the other currency; then multiply the given sum by this
fraction.*

When the reduction is made to English currency, we
multiply by $\frac{5}{24}$, and the result will be pounds and decimals
of a pound. The decimals of a pound are then reduced to
shillings and pence.

Thus, reduce $45.25 cents to English currency.

OPERATION.

$$\$45.25 \times \tfrac{5}{24} = £9.42708 +.$$

Multiply the given sum
by $\frac{5}{24}$, we get £9.42708+.
Bringing now the decimals
of a pound to shillings and
pence, we have £9 8s. 6¼d.
+.

£9.42708
20
———
8.54160
12
———
6.49920
4
———
1.99680

Ans. £9 8s. 6¼d. +.

Q. What fraction of £1 is $1? How are dollars brought to English
currency? What is the general rule for bringing dollars to any other
currency?

195. Since reducing federal currency to that of any
other country is the reverse of the reduction in Case I., it
follows that the multiplier in Case II. will be the fractions
found for the several currencies in Art. 191, with their
terms inverted. That is, to reduce federal money to New
York currency, we multiply by $\frac{4}{10}$; to New England currency, by $\frac{3}{10}$, &c.

Q. When you know the multiplier for reducing foreign to United
States currency, how may you pass back to the foreign currency? Why?
By what do you multiply to pass to New York currency? To New
England? To Georgia? To Pennsylvania? To Canada?

EXAMPLES.

1. Reduce $1000 to the several currencies.

Ans.
$1000=£208 6s. 7d.+ English currency.
$1000=£250 · Canada "
$1000=£300 N. England "
$1000=£400 New York "
$1000=£375 Pennsylvania "
$1000=£233 6s. 7¾d. S. Carolina "
$1000=£500 N. Carolina "

2. Reduce $1750.37 cents to the several currencies.

SIMPLE INTEREST.

196. INTEREST *is the allowance made to the lender of money for the use of the money.*

The *Principal* is the money on which the interest is paid.

The *Amount* is the principal and interest taken together.

Thus, if A borrow $100 from B, and promise to pay him $6 for the use of it for one year, the *interest* would be $6, the *principal* $100, and the *amount* $106; and B would receive at the end of the year $106; that is, the principal $100, and the interest $6.

Q. What is interest? What is the principal? The amount? How do you illustrate this?

197. The *rate* of interest is the interest of $100 for one year. When the rate is $6, it is called 6 per cent. interest; that is, $6 *for the hundred.* Interest at 7, 8, 9, &c. per cent. is an allowance of $7, $8, $9, &c. for the use of $100 for one year. The *legal* interest is the rate established by law. In most of the States it is 6 per cent. In New York it is 7 per cent., and in Louisiana 8 per cent.

Q. What is the rate of interest? When the rate is $6, what is it called? What is meant by 6 per cent. interest? 7 per cent? 8? What is legal interest? What rate is legal interest in most of the States? What is the legal interest in New York? In Louisiana?

198. To find the interest on any given principal for one year, we have only to multiply the principal by the rate per cent. and divide by 100, or cut off two places to the right for decimals.

Thus, the interest on $540 for one year at 6 per cent. will be $\frac{540 \times 6}{100}$ = $32.40.

Since it is evident $100 : $6 :: $540 : $32.40 ; that is, $100 will be to $6, its interest for one year, as $540 is to $32.40, its interest for the same time; and here the fourth term is found by multiplying the third term by 6 and dividing the product by 100.

Having thus found the interest for one year, that for 2, 3, or any number of years, will be ascertained by multiplying the interest of the given principal for one year by 2, 3, &c., since it is plain the interest of a given principal for 2 years will be double that for one year, for 3 years 3 times, &c.

Q. How do you find the interest on a given principal for 1 year? Why is this? How do you find the interest of a given principal for 2 or more years? The reason for this? What is the interest of $200 for 1 year at 6 per cent.? For 2 years? 3 years? 4? What is the interest of $500 for 1 year at 6 per cent.? For 2 years? 3 years? 10 years? What is the interest of $1000 for 1 year at 6 per cent.? For 2 years? 3 years? 10 years?

199. When the rate of interest is 6 per cent. for one year or 12 months, every *two months* will produce $1 for a principal of $100; that is, 1 per cent. upon the principal. Hence, to find the interest of any given principal for any number of years or months, we have the following general

RULE.

Multiply the given principal by half the number of months, and divide the product by 100, or what is the same thing, cut off two places of decimals from the right.

Thus, to find the interest on $500 for 2 years and 5 months at 6 per cent.

OPERATION.

There being 29 months in 2 years and 5 months, the multiplier is 14½, and the result $72.50 is the answer.

$500
14½ = half no. of mos.
————
2000
500
250
————
Ans. $72.50

Again, find the interest of $475.50 for 3 years, 10 months, and 15 days, at 6 per cent.

OPERATION.

In calculating interest, the month is estimated at 30 days; 15 days will therefore be $\frac{1}{2}$ month, and there will be 46$\frac{1}{2}$ months in the given time, and the multiplier will be 23$\frac{1}{4}$.

$475.50
23$\frac{1}{4}$ = half no. of mos.

142650
95100
11887.5

110.55375

Ans. $110.55.3+.

· *Q.* When the rate of interest is 6 per cent., what amount of interest will $100 produce in 2 months? What rate of interest will $1 per $100 be? What is the general rule to find the interest on any given principal for any number of years and months? What would be the multiplier if the time were 4 years and 2 months? How many days are calculated for a month in interest? How is the interest found for any number of days?

EXAMPLES.

1. Find the interest of $80 for 2 years, 6 months, and 13 days, at 6 per cent.

OPERATION.

13 days being $\frac{13}{30}$ of a month, there will be 30$\frac{13}{30}$ months in the given time, the half of which will be 15$\frac{13}{60}$; since to halve a fraction you multiply the denominator by 2, or divide the numerator by 2.

$80
15$\frac{13}{60}$ = $\frac{1}{2}$ no. of mos.

400
80
17$\frac{1}{3}$

Ans. 12.17\frac{1}{3}$

2. Find the interest of $125 for 23 days, at 6 per cent.

OPERATION.

23 days being $\frac{23}{30}$ of a month, the multiplier will be $\frac{23}{60}$, which is the $\frac{1}{2}$ of $\frac{23}{30}$.

$125
$\frac{23}{60}$ = $\frac{1}{2}$ no. of mos.

Ans. 0.47\frac{11}{12}$.

3. Find the interest of $700 for 8 months and 15 days, at 6 per cent.   Ans. $29.75.

4. Find the interest of $327.25 for 9 months and 20 days, at 6 per cent.   Ans. $15.81.7.

5. Find the interest of $1000 for 3 years and 10 days, at 6 per cent: Ans. $181.66.6+.

6. Find the interest of $22.45 for 5 years, 4 months, 11 days, at 6 per cent.   Ans.

7. Find the interest of $1700.39 for 4 months, at 6 per cent.   Ans.

8. Find the interest of $37.21 for 18 months and 3 days, at 6 per cent.   Ans.

9. Find the interest of $59 for 4 years, 7 months, and 10 days, at 6 per cent.   Ans.

10. Find the interest of $74.40 for 9 years, 3 months, at 6 per cent.   Ans.

11. Find the interest of $1111.21 for 3 years, 21 days, at 6 per cent.   Ans.

12. Find the interest of $10,000 for 10 years, at 6 per cent.   Ans.

200. *When the rate of interest is greater or less than 6 per cent., we find the interest of the given principal at 6 per cent.; then take such a part of this interest as the given rate is of 6 per cent.* Thus,

Find the interest of $70 for 1 year and 4 months at 7 per cent.

The interest of $70 for 16 months at 6 per cent. is $5.60. Taking $\frac{7}{6}$ of this result, (since 7 is $\frac{7}{6}$ of 6 per cent.) which is done by multiplying by 7 and dividing by 6, the required interest is $6.53⅓.

OPERATION.

$70

  8 = ⅓ no. of months.
_____

$5.60 = interest at 6 p. ct.

  7
_____

6) 3920
_____

Ans. $6.53⅓ = int. at 7 p. ct.

Again, find the interest of $114 for 2 years and 10 months, at 5 per cent.

The interest at 6 per cent. is $19.38. Since 5 is ⅚ of 6, we take ⅚ of this result for the interest at 5 per cent., which is done by multiplying by 5 and dividing by 6.

$114
17=1½ no. of months.
—————
798
114
—————
$19.38=interest at 6 p. ct.
5
—————
6) 9690
—————

Ans. $16.15=int. at 5 per cent.

Q. How do you find the interest of a given principal when the rate is greater or less than 6 per cent. ? What part of the interest at 6 per cent. would you take to find the interest at 7 per cent.? Why? At 5 per cent. ? Why? At 4 per cent.? At 3? At 2? At 1½ per cent.? At 10 per cent. ? At 8 per cent. ? At 9? What would be the interest of $100 for 1 year at 7 per cent.? At 5? At 3?

EXAMPLES.

1. Find the interest of $12.50 for 12 months, at 7 per per cent. Ans. 87½ cts.

2. Find the interest of $29.43 for 3 years, 11 months, at 8 per cent. Ans. $9.22.13.

3. Find the interest of $79.75 for 25 days, at 10 per cent. Ans.

4. Find the interest of $175.25 for 1 year, 20 days, at 5 per cent. Ans.

5. Find the interest of $200 for 4½ years, at 5½ per cent. Ans. $49.50.

6. Find the interest of $1000 for 2 years, 4 months, at 2½ per cent. Ans. $58.33⅓.

7. Find the interest of $2750 for 22 months, at 1½ per cent. Ans.

8. Find the interest of $947.22 for 2¾ years, at 3 per cent. Ans.

9. Find the interest of $2047.50 for 2 years, 3 months, 10 days, at 4 per cent.   Ans.

10. Find the interest of $850 for 4 years and 19 days, at 4½ per cent.   Ans.

11. Find the interest of $137.25 for 1 year, 8 months, at 5¾ per cent.   Ans. $13.15.3+.

12. Find the interest of $1945.25 for 10 months, 29 days, at 2 per cent.   Ans.

201. *When we have the amount given, as well as the time and rate per cent., we may readily find the principal.*

Thus, what principal at 6 per cent. will in 4 years amount to $248?

The interest of $1 for 4 years at 6 per cent. is 24 cents, and the amount will be $1.24.   Hence,

$$\$1.24 : \$1 :: \$248 : \text{Ans.} = \$\tfrac{248}{1.24} = \$200;$$

that is, the amount of $1 for the given time is to $1, as the given amount is to the required principal; from which we see that the principal is found *by dividing the given amount by the amount of $1 at the given rate and time.*

What principal at 6 per cent. for 1 year and 8 months will amount to $660?

The amount of $1 for 20 months is $1.10; dividing $660 with two ciphers annexed, by $1.10, the answer is $600.

OPERATION.

1,10) $660.00
————————
Ans. = $600

We may readily prove the correctness of this result, for the interest of $600 for 1 year and 8 months at 6 per cent. is $600 × 10 = $60, and $600 + $60 = $660 = the amount.

Q. May you determine the principal when the amount, rate, and time are given? How? What is the amount of $1 for 4 years at 6 per cent.? For 1 year and 8 months? How do you prove the correctness of the rule?

EXAMPLES.

1. What principal at 6 per cent. for 18 months will *amount to $500?*   Ans. $458.71.5+.

2. What principal at 6 per cent. for 2 years and 4 months will amount to $750? Ans. 657.89.4+.

8. What principal at 6 per cent. for 3 years will amount to $244? Ans. $206.77.9+.

4. What principal at 4 per cent. for 2 years will amount to $350.24? Ans.

5. What principal at 3 per cent. for 14 months, 8 days, will amount to $257.25? Ans.

6. What principal at 5 per cent. for 1 year, 4 months, and 6 days, will amount to $1000? Ans.

202. *To find the principal when the time, the rate per cent., and the interest are given, we divide the given interest by the interest of $1 at the given rate and time: the quotient will be the principal required.*

What sum of money put at interest for 1 year and 4 months, at 6 per cent., will produce $50 interest?

The interest of $1 for 16 months being 8 cents, the interest of $1 will be to $1, as the given interest $50 is to the required principal.

OPERATION.

cts. cts. cts. cts.

$8 : 100 :: 5000 : \frac{5000 \times 100}{8} = \$625.$

Q. How do you find the principal, when the time, the rate per cent. and the interest are known? How do you explain this?

EXAMPLES.

1. What principal placed at interest for 2 years and 10 months at 6 per cent. will produce $47.50 interest? Ans.

2. What principal placed at interest for 3¾ years at 6 per cent. will produce $100 interest? Ans.

8. What principal placed at interest for 1 year, 8 months, and 10 days, at 6 per cent., will produce $1500? Ans.

4. What principal placed at interest for 1 year at 6 per cent. will produce $1000 interest? Ans.

203. *When the principal, interest, and time are known, we may find the rate per cent. by dividing the given interest by the interest of the given sum at 1 per cent. for the given time.*

Thus, if \$500 placed at interest for 1 year produce \$30, what is the rate per cent. ?

| \$500 at interest for 1 year | OPERATION. |

\$500 at interest for 1 year
at 1 per cent. will produce
\$5. Hence, the interest of
\$500 for 1 year at 1 per
cent. is to 1 per cent. as the

OPERATION.

$$\$5 : \$1 :: \$30 : \$\tfrac{30}{5} = \$6.$$

Ans. 6 per cent.

given interest \$30 is to 6, the required rate per cent.

Q. How do you find the rate per cent. when the principal, interest, and time are known ? Explain the reason of this rule.

EXAMPLES.

1. If \$700 placed at interest for 1 year produce \$14, what is the rate per cent.?   Ans. 2 per ct.

2. If the interest of \$1500 for 2 years amount to \$90, what is the rate per cent. ?   Ans. 3 per ct.

3. If the interest of \$2000 for 3 years and 6 months amount to \$350, what is the rate per cent. ?

Ans.

204. *When the principal, rate per cent., and interest are given, we may find the time by dividing the given interest by the interest of the principal for 1 year at the given rate.* Thus,

If the interest of \$500 at 6 per cent. amount to \$45, how long was the principal at interest ?

The interest of the   OPERATION.
principal \$500 for 1
year at 6 per cent. is

$$\$30 : 1 \text{ yr.} :: \$45 : \tfrac{45}{30} \text{ yr.} = 1\tfrac{1}{2} \text{ yr.}$$

\$30 ; hence the interest \$30 will be to its time, 1 year, as the given interest \$45 is to 1½ years, the required time.

Q. When the principal, rate per cent. and interest are given, how do *you find the time ?* Explain the reason of this rule.

EXAMPLES.

1. What time must $1000 be at interest at 6 per cent. to produce $120 ?   Ans. 2 years.

2. How long must $500 be at interest at 6 per cent. to produce a sum equal to the principal?   Ans. 16⅔ years.

3. How long must $500 be at interest at 7 per cent. to produce a sum equal to the principal?   Ans.

4. How long must $1500 be at interest at 3 per cent. to produce $100?   Ans.

COMPOUND INTEREST.

205. When the interest on a given principal becomes due and is not paid, but is added to the principal, the interest calculated upon the amount as a new principal is called COMPOUND INTEREST.

Thus, if A borrow $100 from B for one year at 6 per cent., A will owe B $106 at the end of the year. If he pays neither the principal nor interest at this time, but keeps the amount for another year, the interest calculated on $106 for one year will be compound interest.

The method of calculating compound interest for any period of years consists in *finding the interest on the given principal for the first year at the given rate; then adding this interest to the principal for a new principal, calculating the interest on this new principal for another year, and adding the principal as before. Proceed in this way for each year, and from the last result subtract the given principal, we shall have the compound interest required.*

EXAMPLES.

1. Find the compound interest of $500 for 3 years at 6 per cent.

I

OPERATION.

$500=given principal.
    6
_____

30.00=interest for 1st year.
500
_____

530.00=am't. for 1st yr. or principal for 2d yr.
    6
_____

31.8000=interest for 2d year.
530
_____

561.8000=principal for 3d year.
    6
_____

33.708000=interest for 3d year.
561.8000 .
_____

595.508000=amount for 3 years.
500   =given principal deducted.
_____

Ans.=$95.50.8.000=compound interest for 3 years.

2. Find the compound interest of $300 for 4 years, at 7 per cent.   Ans.

3. Find the compound interest of $125 for 3 years and 8 months, at 5 per cent.   Ans.

4. Find the compound interest of $325 for 1 year, 10 months, and 4 days, at 6 per cent.   Ans.

5. What is the difference between the compound interest and simple interest of $1000 for 2 years and 6 months, at 6 per cent.?   Ans.

6. Find the compound interest of $1525 for 8 years and 10 months, at 4 per cent.   Ans.

7. What is the difference between the simple interest and compound interest of $1800 for 5 years, at 5 per cent.?   Ans.

8. Find the compound interest of $275 for 3 years, 4 months, and 10 days, at 6 per cent.

OPERATION.

$275
6
___
16.50
275
___
291.50
6
___
17.4900
291.50
___
308.9900
6
___
18.539400
308.99
___
327.529400
275
___
52.529400 comp. int. for 3 yr.
7.096470+ " 4 m. & 10 dy.
___
Ans. $59.625870

When it is required to find the compound interest for yrs., mos., and days, it is found most convenient to find the compound interest for the years first, and then calculate the interest for the months and days of the amount for the years, and add this interest to the compound interest for the years; the sum will be the interest required.

In this example we find the compound interest for 3 years; and to this add $7.096470, which is the interest on the amount $327.5294 for 3 years, for the compound interest required. We multiply by $2\frac{1}{6}$, half the number of months $4\frac{1}{3}$.

$327.5294
$2\frac{1}{6}$
___
6550588
545882+
___
Ans. $7.096470

Q. What is compound interest? Illustrate it. How do you find compound interest for years? For years, months, and days?

ANNUITIES.

206. An Annuity is a sum of money, payable yearly, *for a certain* period of years, or for ever. Pensions allowed to soldiers, &c. for public services, come under the head of annuities; so do the incomes from the rent of lands, houses, &c.

When an annuity is not paid at the time it becomes due, interest is allowed upon it at the lawful rate.

Annuities are calculated either at simple or compound interest.

Q. What is an annuity? What kind of incomes are comprehended in annuities? When the annuity becomes due, is interest allowed? At what rate? Are annuities calculated at simple or compound interest?

207. The *amount of an annuity* is the sum of all the annuities remaining unpaid, with the interest on each for the time due.

To find the amount of an annuity at simple interest, we calculate as in Art. 199, the interest of each annuity for the time due, and then its amount; the sum of all the amounts will be the amount required.

If the annuity be at compound interest, the rule in Art. 205 will give the interest.

What is the amount of an annuity of $300 for 6 years, at 6 per cent. simple interest?

OPERATION.

Amount of $300 for 5 years = $390
" " $300 " 4 " = $372
" " $300 " 3 " = $354
" " $300 " 2 " = $336
" " $300 " 1 " = $318
6th year's annuity paid when due, = $300
_____
Ans. $2070

If the annuity had been at compound interest, we should find the amount as follows:

OPERATION.

Amount of $300 for 5 years = $401.46 + (Art. 205)
" " $300 " 4 " = $378.74 +
" " $300 " 3 " = $357.30 +
" " $300 " 2 " = $337.08
" " $300 " 1 " = $318.00
6th year's annuity paid when due, = $300.00
_____
Ans. $2092.58 +

Q. What is meant by the amount of an annuity? How is the amount of an annuity found at simple interest? At compound interest? If the annuity run for 6 years, is the interest calculated for the whole time? Why not? When does the 6th annuity become due? If paid when due, should interest be added?

EXAMPLES.

1. What is the amount of an annuity of $60 for 7 years, at 6 per cent. simple interest?   Ans.

2. What is the amount of an annuity of $75 for 3 years, at 6 per cent. compound interest?   Ans.

208. The present value of an annuity which is to continue for a term of years may be readily deduced from Art. 201; since to determine the present value reduces itself to find the principal, which at the given interest and time would amount to the given annuity.

Thus, to find the present value of an annuity of $400, to continue 3 years at simple interest (Art. 201).

Present value of $400 payable at
the end of the 1st year, $= \$\frac{400}{1.06} = \$377.35 +$
Present value of $400 payable at
the end of the 2d year, $= \$\frac{400}{1.12} = \$357.14 +$
Present value of $400 payable at
the end of the 3d year, $= \$\frac{400}{1.18} = \$338.98 +$

Present value required, $= \$1073.47 +$

The present value of each year's annuity being determined by dividing the annuity by the *amount* of $1 for the given time and rate. The sum of these results will give the present value sought.

Q. How may the present value of an annuity be ascertained? To what does the question reduce itself? How is the present value of each year's annuity found? How is the required present value found?

EXAMPLES.

1. What is the present value of an annuity of $60 to continue 5 years, at 6 per cent. simple interest?   Ans.

2. What is the present value of an annuity of $100, to continue 4 years, at 6 per cent. simple interest?

Ans.

COMMISSION AND INSURANCE.

209. COMMISSION is an allowance made to an agent for buying or selling goods, &c., or disbursing money, and is usually a certain *per cent.* upon the value of the articles bought or sold, or upon the money disbursed.

INSURANCE is an allowance made to an individual or company for insuring property from loss by fire, shipwreck, &c., and is also a certain per cent. on the amount insured.

The allowance in either case is found by multiplying the given sum by the rate per cent., and dividing the product by 100, *or cutting off two figures to the right.*

Q. What is commission? How is it estimated? What is insurance? How is it estimated? What is the rule for finding the allowance for commission or insurance?

EXAMPLES.

1. What is the commission on $1200 at $2\frac{1}{2}$ per cent.?
Ans. $30.

2. What would be the commission for selling $15,000 worth of flour, at 4 per cent.?   Ans. $600.

3. What would be the insurance of a house, valued at $4500, for 1 year, at $\frac{1}{4}$ per cent.?   Ans. $11.25.

4. What is the insurance on $20,000, at $\frac{3}{4}$ per cent.?
Ans. $150.

5. What is the commission for selling 100 barrels of pork, valued at 10\frac{1}{2}$ per barrel, at $2\frac{1}{2}$ per cent.?   Ans.

6. What is the commission on the sale of 100 bales of cotton, valued at $50 per bale, at 4 per cent.?
Ans.

DISCOUNT.

210. Discount is an allowance made for the payment of a debt before it becomes due, at a certain rate per cent. for the time allowed.

It is evident, that if we ascertain the present value of the debt at the given rate of interest, the discount will be found by subtracting the present value from the given debt.

To find the present value reduces itself to find the principal, when the amount, the rate per cent., and the time are given, as in Art. 201.

Q. What is discount? How is discount found? How is the present value of the debt ascertained? How do you find the principal, when you know the amount, the rate per cent., and the time?

EXAMPLES.

1. What is the discount on $1000, due 1 year and 8 months hence, at 6 per cent. per annum?

OPERATION.

$$\$\tfrac{1000}{1.10} = \$909.09 + = \text{present value of } \$1000.$$

$$\$1000 - \$909.09 = \$90.91 = \text{discount}.$$

We find the present value by dividing the given sum by $1.10, the amount of $1, for 1 year and 8 months, Art. 201. The present value subtracted from $1000, gives the discount, which is $90.91.

2. What is the discount on $600, due 2 years hence, at 6 per cent.?   Ans. $64.29.

3. What is the discount on $550, due 1 year and 8 months hence, at 6 per cent.?   Ans. $50.

4. What is the discount on $2200, due 1 year and 4 months hence, at 6 per cent.?   Ans.

5. What is the discount on $20,000, due 5 years hence, at 6 per cent.?   Ans.

PROFIT AND LOSS.

211. PROFIT AND LOSS is a rule by which we ascertain the profit or loss upon the sale or purchase of goods; and it also determines at what rates goods must be sold, to gain or lose a certain per cent.

EXAMPLES.

1. Bought 150 barrels of flour, at $5¼ per barrel, and sold them at $5.37½: what was the gain on the sale?

OPERATION.

$150 \times \$5.37\frac{1}{2} = \$806.25 =$ amount of sale.
$150 \times \$5\frac{1}{4} \quad = \$787.50 =$ cost.
$$\$18.75 = \text{profit.}$$

Or thus:

$\$5.37\frac{1}{2} - \$5\frac{1}{4} = 12\frac{1}{2}$ cts. profit per barrel.
$150 \times 12\frac{1}{2}$ cts. $= \$18.75$ total profit.

We first find the total cost of the flour by multiplying the number of barrels by the cost of each, and subtract this result from the total value of the sales, ascertained in the same manner. Or, we may find the profit on one barrel, and multiply the total number of barrels by the profit on one barrel, for the total profit.

2. Bought cloth at $3.50 per yard, and sold it again at $4.20 per yard: what is the gain per cent.?

OPERATION.

$\$4.20 - \$3.50 = 70$ cts. gain per yard.
$3.50 : 100 :: 70 : \frac{70 \times 100}{350} = 20$ per ct.
that is, 20 cts. gain in 100 cents.

Having found the gain on the cost of one yard to be 70 cents, to find the gain per cent., that is, per 100 cents, it is evident, that if 70 cents be the gain on $3.50, the gain on 100 cents will bear the same proportion to 70 cents that *100 does* to $3.50. Hence, by dividing the profit or loss *per yard*, by the cost per yard, the rate per cent. is found.

Q. What is profit and loss? How do you find the profit or loss upon the sale of any number of barrels of flour? How would you find the rate per cent. of the profit or loss? What do you mean by the rate per cent.?

2. Bought 100 bales of cloth of 50 pieces each, each piece containing 20 yards, at \$2.75 per yard. I sold the whole at \$3.08 per yard. What was the profit on the sale, and the rate per cent.?   Ans. $\left\{\begin{array}{l}\text{Total profit, \$33000.} \\ \text{Rate per cent.}=12.\end{array}\right.$

3. Bought 1000 barrels of flour at \$4.25, and sold 500 barrels at \$5¼, and \$500 at \$5⅜ per barrel: what was the gain per cent. on the whole transaction?

Ans. $22\frac{1}{17}$ per cent.

4. Bought 20 hhds. of sugar of 20 cwt. each, at $4\frac{1}{4}$ cents per lb. I then paid 25 cents cooperage per hhd., and lost by leakage 15 lbs. per hhd. The whole was sold at \$4.20 per cwt.: did I gain or lose by the transaction, and what was the total loss or gain, and the rate per cent.?

Ans.

EQUATION OF PAYMENTS.

212. EQUATION OF PAYMENTS *shows the method of finding the mean time of payment of several sums, due at different times, so that there will be no loss or gain of interest.*

It is plain, that \$1 will require 12 months to gain as much interest as \$2 will in 6 months; that is, $6 \times 2 = 12$.

Again, \$1 will require 72 months to gain as much interest as \$6 would in 12 months; that is, $6 \times 12 = 72$.

Again, \$1 will require 100 months to gain as much interest as \$5 would in 20 months; that is, $5 \times 20 = 200$.

Hence, if I owed \$5 to be paid in 4 months, and \$10 to be paid in 12 months, but wished to pay both sums at once, so as not to lose or gain any interest—

The int. of \$5 for 4 mos. $=$ int. of \$1 for $5 \times 4 = 20$ mos.
"   "   " \$10 " 12 " $=$ "   " \$1 " $10 \times 12 = 120$ "

Hence, the int. of \$15 to times of pay't.$=$int. of \$1 for 140 "

12

But if $1 produce a certain interest in 140 months, $15 will produce the same interest in $\frac{140}{15}$ months.

Hence, we have the following

RULE.

Multiply each payment by the time, and divide the sum of these separate products by the sum of the payments; the result will be the answer.

EXAMPLES.

1. I owe $1500, payable as follows: $500 in 6 months; $500 in 9 months; and $500 in 12 months. When must I pay the whole to gain and lose no interest?

In this example, we multiply each payment by its time, and divide 13500, the sum of the products, by 1500 the sum of the payments. The answer is 9 months.

$$
\begin{aligned}
&\text{OPERATION.}\\
500 \times 6 &= 3000\\
500 \times 9 &= 4500\\
500 \times 12 &= 6000\\
\hline
1500 \quad 15,&00\,)\,135,00\\
\hline
&\text{Ans.} = 9 \text{ months.}
\end{aligned}
$$

2. I owe $100, payable as follows: ¼ in 4 months, ⅓ of the remainder in 6 months, the balance in 12 months. What is the mean time of payment?　　　　Ans.

3. Due $3500, payable as follows: $1000 cash in hand, $1000 in 6 months, $1000 in 12 months, $500 in 18 months. What is the mean time of payment?　　　　Ans.

Q. What is equation of payments? How long will it take $1 to gain as much interest as $2 will in 6 months? As $3 will in 8 months? As $5 will in 10 months? In general, how long will it take $1 to gain the same interest as a given sum will in a given time? If $1 produce a given interest in a given time, how long will $2 take to produce the same interest? $3? $5? Any number of dollars? What is the rule for finding the mean time of payment?

APPLICATIONS UPON THE FOREGOING RULES.

1. Find the total amount of the debts of the following States in Federal Currency, their respective indebtedness in *1789*, when the United States government commenced operations, being as stated below in continental currency.

	Continental value.				Value in Federal
	£	s.	d.		Currency.
1. *Massachusetts*,	1,568,040	7	9	=	$5,226,801.29
2. *Connecticut*, ..	585,352	0	0	=	1,951,173.33+
3. *New York*, ...	467,030	2	0	=	1,167,575.25
4. *New Jersey*, ..	295,755	4	11	=	788,680.65+
5. *Virginia*, ...	1,104,222	18	2	=	
6. *South Carolina*,	1,256,787	9	7	=	

Ans. =$18,201,205.59+

Q. When did the United States government go into operation? How was money estimated at that time? How many shillings to the dollar in Massachusetts? Connecticut? New York? New Jersey? Virginia? South Carolina? How would you convert pounds, shillings, and pence, Massachusetts currency, to federal currency? How would you convert Connecticut currency to federal currency? New York currency to federal currency? New Jersey? Virginia? South Carolina? What was the total indebtedness of the States in 1789?

2. What was the total amount of the foreign public debt of the United States in 1789, in federal currency, there being due on loans to France 34,000,000 livres, and to Holland 9,000,000 florins? Ans. $9,890,000.

Q. To what countries was the foreign debt of the United States due in 1789? What is the value of the French livre? How are French livres converted into federal currency? What is the value of the Dutch florin? How converted into federal currency?

3. The foreign and domestic debt of the United States in 1789, including arrears of interest, was as follows:

Foreign debt, ... $11,710,378.62
Domestic " 42,414,085.94

What would the interest on the foreign debt and on the domestic debt amount to, per annum, at 4 per cent., and what annual income was required to pay the interest as it fell due?

Ans. {
Interest on foreign debt = $468,415.14
" " domestic " 1,696,563.43
Annual income required, = $2,164,978.57
}

Q. What was the amount of the United States foreign debt in 1789? Domestic debt? What annual income was required to pay the interest at 4 per cent.?

4. What was the total value of the coinage of the United States Mint at Philadelphia in 1842, the number of pieces of each denomination being as follows:

	No. of pieces.	Value.
Eagles, ...	81,507	= $815,070.00
½ Eagles, ...	27,578	= $137,890.00
¼ Eagles, ...	2,823	= $
Dollars, ...	184,618	= $
½ Dollars, ...	2,012,764	= $
¼ Dollars, ...	88,000	= $
Dimes, ...	1,887,500	= $
½ Dimes, ...	815,000	= $
Cents, ...	2,383,390	= $

Ans. $2,426,351.40

5. What was the amount of the funded debt contracted by the United States government during the war with England, which commenced in 1812 and ended in 1815, the whole amount being embraced in the following loans:

1. $7,860,500 at par,* .... = $ 7,860,500.00
2. $16,000,000, obtained at the rate of
$88 in cash for $100 in stock, = $18,181,818.18
3. $7,500,000, obtained at the rate of
$88.25 in cash for $100 in stock, = $ 8,498,583.56
4. $12,292,868.90 at 80 per ct. stock, = $15,366,111.12
    140,810.00 at 85   "   " = $ 165,656.82
    43,222.22 at 90¾   "   " = $ 47,627.79
    74,590.75 at 90¼   "   " = $ 82,420.72
5. $7,924,219.59 at 95   "   " = $ 8,341,283.77
    1,047,846.30 at 96½   "   " = $
    32,978.49 at 97   "   " = $
    275,000.00 at 98   "   " = $
    4,000.00 at par, = $
6. $9,268,949.00 at par, = $

Ans. $63,217,414.72

* *Note.*—When $100 of stock sells for $100, the stock is at *par;* when for less than $100, it is *below par;* when for more than $100, it is *above par.*

Q. How did the United States government obtain the funds for carrying on its war with England? When did the last war begin? When end? How much money was borrowed? What security did the United States give? (Its own stock.) Was it generally above or below par? When is stock par? When below par? When above par? How much stock at 95 per cent. must be given for $100 cash? How much for $1000 cash? How much stock at 90 per cent. must be given for $100 cash? How much at 80? What is $1000 stock worth at par? At 5 per cent. above par? At 10 per cent. above par? At 20 per cent.?

6. By a treaty with the French in 1831, to indemnify the United States for spoliations upon their commerce, the French agreed to pay 25,000,000 of francs, in 6 annual instalments, bearing interest at 4 per cent.; what was the amount of each payment in federal currency?   Ans.

7. The last United States Bank was chartered by Congress in 1816, for 20 years, with a capital of 35,000,000 of dollars, the United States government holding 70,000 shares at $100 each; what advance per cent. would there be in the stock to make the capital of the government $8,000,000?   Ans.

8. In 1829, a dividend of 3½ per cent. was declared upon the capital stock of the United States Bank, for the previous 6 months; what amount of dividend was paid to the United States, its total amount in stock being $7,000,000?

Ans. $245,000.

9. By the law of Congress establishing the last Bank of the United States, a bonus of $1,500,000 was stipulated to be paid to the United States; what per cent. of the capital stock did the sum amount to?   Ans.

10. What distance will sound travel in 1 minute through the following media, its rate per *second* being as follows:

Through air, ... 1143 ft. per sec. . Ans. 12.7 miles per m.
"   water, ... 4900   "   "   Ans.
"   cast iron, 11090 "   "   Ans.
"   steel, .. 17000   "   "   Ans.
"   glass, . 18000   "   "   Ans.

Q. Through what media does sound travel most rapidly?

11. The best bell-metal contains 80 per cent. of copper, and 20 per cent. of tin: how many pounds of each substance would be in the following bells, supposing the bells to be made in the above proportion?

	Weight.	Copper.	Tin.
Empress Ann's, Moscow,	432,000 lbs.	Ans. 345,600 lbs.	86,400 lbs.
Borts Godemits, "	288,000	Ans. 230,400	57,600
St. Ivan's, "	142,000	Ans. 113,600	28,400
Great Bell, Pekin,	120,000	Ans. 96,000	24,000
Novogorod great bell,	70,000	Ans. 56,000	14,000
Great Tom, Oxford,	18,000	Ans. 14,400	3,600

Q. What metals compose bell-metal? What proportion of each? What are some of the largest bells in the world?

12. The steam-ships British Queen and President are stated to have cost £90,715 each. The total receipts of the British Queen for 9 voyages amounted to £82,001 2*s.* 6*d.*, and the total expense £70,691. The total receipts of the President for 9 voyages were £25,234 6*s.* 7*d.*, and the expenses £21,833 3*s.* 7*d.* What was the average profit per voyage in federal currency, and what rate per cent. was the profit upon the first cost? Ans.

13. The first voyages from Great Britain to New York by steam were made simultaneously by the steam-ship "Sirius" from Cork, and the Great Western from Bristol, in 1838. The former travelled 3,300 miles, and performed the voyage in 19 days. The Great Western travelled 3,223 miles, and performed the voyage in 15 days 5 hours. What was the average rate per hour of each vessel?

Ans. $\begin{cases} \text{Sirius' rate per hour, 7.2 + miles.} \\ \text{G. W.'s } " \quad " \quad " \quad 8.8 + \text{miles.} \end{cases}$

14. The total amount of debt of Great Britain in 1836, was £787,638,816 8*s.* 9¼*d.* : what is the annual charge to that government, estimating the interest at 3 per cent. in federal money? Ans.

15. By the act of Congress of 1835, the standard of the gold and silver coins of the United States was fixed as fol lows. That of 1000 parts by weight, 900 shall be pure metal and 100 parts alloy; the alloy of the silver coins being *copper; that* of the gold coins, equal parts of silver and *copper. The* dollar weighs 412½ grains; ½ dollar, 206¼;

¼ dollar, 103¼ ; the dime, 41¼ ; and the ½ dime, 20⅝ grains. The eagle weighs 258 grains ; ½ eagle, 129 ; and the ¼ eagle, 64¼ grains. The copper coins, composed of pure copper, as follows : the cent weighs 168, and the ½ cent 84 grains. How much gold, silver, and copper, by weight, were used in the coinage at the United States Mint, Philadelphia, in 1842, the number of pieces of each kind being as follows :

Kinds.	Pieces.	Gold.	Silver.	Copper.
Eagles,	81,507			
½ "	27,578			
¼ "	2,823			
Dollars,	184,618			
½ "	2,012,764			
¼ "	88,000			
Dimes,	1,887,500			
½ "	815,000			
Cents,	2,383,390			

Ans. 3870.0 lbs. 93188.6 lbs. 80060.9 lbs.

Q. What are the coins of the United States ? By whom is their weight fixed ? What metals compose the gold coins ? The silver ? How much alloy to the 1000 parts ? What is the weight of the eagle ? Dollar ? Cent ? Is the weight determined by the pound avoirdupois or troy ?

DUODECIMALS.

213. DUODECIMALS are numbers in which the unit has been divided into *twelfths*, or some multiple of twelve. The name is derived from the Latin word *duodecim*, which means *twelve*.

As usually applied, the unit which is thus divided is the *foot*.

When 1 foot is divided into 12 equal parts, each of these parts is called *primes*, or more commonly *inches*, and is denoted by the sign '.

The inch or prime being again divided into twelve equal parts, each of these parts is called *seconds*, and is de by the sign ".

In like manner, the second being divided into twelve other equal parts, the name of *thirds* is given to them, and they are denoted by the sign ''', and so on.

The number 1 ft,, 5', 10'', 6''', is read, 1 foot, 5 inches, 10 seconds, and 6 thirds.

Q. What are duodecimals? From what is the name derived? To what unit is this division usually applied? When the foot is divided into twelve equal parts, what is each of these parts called? How denoted? When the inch is divided into twelfths, what is each of the equal parts called? How denoted? What name is given to the twelve equal parts into which the seconds are divided? How denoted? How would you read the expression 5 ft. 2' 3'' 4'''? How would you write the number 4 feet, 3 inches, 4 seconds, 2 thirds?

214. From the manner in which these divisions of the foot are formed, it follows that

1 inch $= \frac{1}{12}$ of a foot.

1 second$= \frac{1}{12}$ of 1'$= \frac{1}{12}$ of $\frac{1}{12}$ of a foot$= \frac{1}{144}$ of a foot.

1 third$= \frac{1}{12}$ of 1''$= \frac{1}{12}$ of $\frac{1}{12}$ of $\frac{1}{12}$ of a foot$= \frac{1}{1728}$ of a foot.

Q. What part of 1 foot is 1 inch? What part of 1 inch is 1''? What part of 1 foot is 1''? What part of 1'' is 1'''? What part of 1 foot is 1'''?

215. When 1 foot is multiplied by 1 foot, the product is 1 *square* foot.

When 1 foot is multiplied by 1 inch, since 1'$= \frac{1}{12}$ of a foot, the product is $1 \times \frac{1}{12} = \frac{1}{12}$ of a foot$=1'$.

When 1 inch is multiplied by 1 inch, since 1'$= \frac{1}{12}$ of a foot, the product is $\frac{1}{12} \times \frac{1}{12} = \frac{1}{144}$ of a foot$=1''$.

When 1 inch is multiplied by 1'', since 1'$= \frac{1}{12}$ of a foot, and 1''$= \frac{1}{144}$ of a foot, the product is $\frac{1}{12} \times \frac{1}{144}$ of a foot$= \frac{1}{1728}$ of a foot$=1'''$.

Hence we conclude that—

Feet multiplied by feet give square feet in the product.

Feet multiplied by inches give inches in the product.

Inches multiplied by inches give seconds in the product.

Inches multiplied by seconds give thirds in the product.

Q. When feet are multiplied by feet, what will the product be ? Why? Feet by inches ? Why? Inches by inches ? Why? Inches by seconds ? Why?

ADDITION AND SUBTRACTION OF DUODECIMALS.

216. Addition and subtraction of duodecimals are performed like the corresponding operations in compound numbers.

EXAMPLES.

1. Add 2 ft. 3' 4" to 4 ft. 10' 11".

Commencing on the right, 11" and 4" make 15"=1' and 3" over; set down 3", and carry 1' to the next unit, and so on throughout.

```
        OPERATION.
       ft.    '    "
        2    3    4
        4   10   11
       ─────────────
  Ans.  7    2    3
```

2. From 11 ft. 3' 1" 4''' take 5 ft. 6', 2", 3'''.

In this example, 3''' from 4''' gives 1'''; 2" from 1" we cannot, but borrowing 12", 2" from 13" gives 11"; carrying 1' to the next higher unit, and proceeding as in subtraction of compound numbers, the result is 5 ft. 6' 11" 1'''.

```
        OPERATION.
       ft.   '    "   '''
       11   3    1    4
        5   8    2    3
      ──────────────────
  Ans. 5    6   11    1
```

13. Add 4 ft. 0' 2" 3''' and 5 ft. 9' 3" 9''' together.
Ans. 9 ft. 9' 6" 0'''.

4. From 8 ft. 0' 0" 2''' take 2 ft. 1' 0" 3'''.
Ans. 5 ft. 10' 11" 11'''.

Q. How are addition and subtraction of duodecimals performed ?

MULTIPLICATION OF DUODECIMALS.

217. Multiplication of Duodecimals is frequently used to determine the contents of solid bodies, such as timber and the like, when the dimensions of its sides are known in feet and parts of a foot, since the number of solid feet will be equal to the product of the length, breadth, and thickness.

1. Find the solid contents of a stick of timber, w measures 13 ft. 2' in length, 4 ft. 9' in breadth, and 2 in thickness.

Arranging the numbers, so that feet shall fall under feet, inches under inches, &c., we multiply the lowest unit in the multiplicand by the highest unit in the multiplier, that is 9' by 4 ft. Since inches multiplied by feet give inches, (Art. 215), the product is 36'=3 ft.; we set down 0' and carry 3 to the feet; then multiplying 13 feet by 4 ft. and adding in the 3 ft. carried, the first partial product is 55 ft. Multiplying now by 3', since inches multiplied by inches give seconds, (Art. 215), 9' multiplied by 3' will

OPERATION.		
ft.	'	
13	9	
4	3'	
55	0'	
3	5'	3''
58	5'	3''
2	4'	
116	10'	6''
19	5	9
Ans. 136	4''	3''

be 27''=2' and 3'' over; set down 3'' and carry 2' : 3 t 13 are 39 and 2 are 41'=3 ft. and 5' over; set down the column of inches, and carry 3 ft. which are set (in the column of feet. Multiplying now this entire pro by the thickness of the timber, which is 2 ft. 4', and serving that feet multiplied by feet give feet, feet by in inches, inches by inches, seconds, and inches by sec thirds, (Art. 215), the solid content of the timber is 1: 4' 3''.

2. How many solid feet in a log measuring 20 ft. 6' 3 ft. 2' broad, and 3 ft. 5' thick? Ans.

3. Find the solid contents of a box, which measures 9' in depth, 4 ft. 5' in width, 2 ft. 8' in thickness.
Ans.

4. Find the solid content of a piece of timber which sures 45 ft. 11' in length, 2 ft. 7' in width, and 3 ft. 6' t
Ans.

Q. What use is made of multiplication of duodecimals? How content of a solid body determined?

FORMATION OF POWERS.

218. THE POWER of a number is the result obtained by multiplying it any number of times by itself.

Any number is the *first power* of itself.

If a number be multiplied by itself, the product is called the *second power* or *square* of that number.

If a number be multiplied by itself, and the resulting product be again multiplied by the number, the last product is called the *third power* or *cube* of the number.

In like manner, the *fourth, fifth,* &c. *powers* of the number will be the product resulting from the multiplication of the number by itself; three times for the fourth power, four times for the fifth power, &c.; the number which designates the power being always greater by 1 than the number of times the number is multiplied by itself.

The number which designates the power is called the *index* or *exponent* of the power, and is written on the right, and a little above the given quantity.

Thus,

$$3 = 3^1 = \quad 3 \text{ is the 1st power of 3;}$$
$$3 \times 3 = 3^2 = \quad 9 \text{ is the 2d power of 3;}$$
$$3 \times 3 \times 3 = 3^3 = \quad 27 \text{ is the 3d power of 3;}$$
$$3 \times 3 \times 3 \times 3 = 3^4 = \quad 81 \text{ is the 4th power of 3;}$$
$$3 \times 3 \times 3 \times 3 \times 3 = 3^5 = 243 \text{ is the 5th power of 3.}$$

Hence the formation of the power of a number consists in finding the product arising from its being multiplied by itself a certain number of times.

Q. What is the power of a number? What is the 1st power of a number? The 2d power? What else is the 2d power of a number called? What is the 3d power of a number? Cube? What is the 4th power of a number? 5th power? How does the number which designates the power of a quantity compare with the number of times it is multiplied by itself? What is the exponent or index of a power? How is it written? In the expression 3^1, which is the index? What is the power? Which is the exponent in 3^2? What does it denote? To what is 3^2 equal? What does 3^4 mean? 3^5? What is the value of 3^4? 3^5? In what does the formation of powers consist?

EXAMPLES.

1. What is the 4th power of 25?

Since the 4th power is required, the given number is multiplied by itself 3 times.

OPERATION.

$25 \times 25 \times 25 \times 25 = 390625.$

2. Find the square of 36.   Ans. 1296.

3. Find the cube of 25.   Ans. 15625.

4. Find the 4th power of 100.   Ans. 100000000.

5. Find the 5th power of 30.   Ans. 24300000.

6. Find the 10th power of 10.   Ans.

7. Find the square of $\frac{1}{2}$.   Ans. $\frac{1}{4}$.

8. Find the square of .5.   Ans. .25.

9. Find the cube of $\frac{3}{4}$.   Ans.

10. Find the cube of .75.   Ans.

11. Find the square of $3\frac{1}{3}$.   Ans. $\frac{100}{9}$.

12. Find the cube of $5\frac{1}{4}$.   Ans.

13. Find the 7th power of 1.   Ans.

14. Find the 4th power of .01.   Ans.

15. Find the 3d power of $\frac{1}{2}$ of $\frac{3}{4}$.   Ans.

EXTRACTION OF ROOTS.

219. A Root is that quantity which, multiplied by itself a certain number of times, will produce a given power. Thus, 2 is the root of 4, since $2 \times 2 = 4$.

When a number multiplied by itself once produces the given power, it is called the *square*, or 2d *root*. Thus, 2 is the square root of 4.

The cube root of a quantity is that number which, multiplied by itself twice, will produce the given quantity.

2 is the cube root of 8, since $2 \times 2 \times 2 = 8$. 3 is the cube root of 27, since $3 \times 3 \times 3 = 27$.

Extraction of Roots consists in finding the root of a given power.

To denote that the square root is to be extracted, we use the sign $\sqrt{}$, which is placed over the number. Thus, $\sqrt{4}$, is read square root of 4, and indicates that the square root of 4 is to be extracted.

When the cube root is to be extracted, we use the sign $\sqrt[3]{}$; thus, $\sqrt[3]{9}$ is read cube root of 9, and indicates that the cube root of 9 is to be extracted.

In like manner, $\sqrt[4]{}$, $\sqrt[5]{}$, &c., represent that the 4th and 5th, &c. roots are to be extracted. The number placed over the sign is called the *index* of the root, and indicates what root is to be extracted. When no index is expressed, 2 is understood.

A number is said to be a *perfect power*, when its exact root can be taken.

The following numbers in the 1st line are *perfect squares* of the corresponding numbers below them:

1	4	9	16	25	36	49	64	81	100 squares.
1	2	3	4	5	6	7	8	9	10 roots.

The following numbers in the 1st line are *perfect cubes* of the corresponding numbers below them:

1	8	27	64	125	216	343	512	729 cubes.
1	2	3	4	5	6	7	8	9 roots.

Q. What is a root? What is the square root of a quantity? Cube root? 4th root? 5th root? What is the square root of 4? Why? Of 9? Why? What is the cube root of 27? Why? Of 64? Why? In what does the extraction of roots consist? How do you denote that the square root has to be extracted? Cube root? 4th root? 5th root? Write upon your slate the cube root of 27. Square root of 64. 4 times the square root of 9. 4th root of 100. What is the index of a root? What does it indicate? When the index is not expressed, what is understood? When is a number said to be a perfect power? What is the square of 2? Of the 1st 10 numbers? What is the square root of 1? 4? 9? 16? 25? 36? 49? 64? 81? 100? What is the cube root of 1? 8? 27? 64? 125? 216? 343? 729? 1000?

EXTRACTION OF THE SQUARE ROOT.

220. EXTRACTION of the square root of a number consists in finding a number which, multiplied by itself once, will produce the given number.

We have seen (Art. 219) that there are only nine perfect squares which can be expressed by one or two figures, viz.

1, 4, 9, 16, 25, 36, 49, 64, 81,

the square roots of which are the first nine numbers, 1, 2, 3, 4, 5, 6, 7, 8, 9. *The square root of any number express-ed by one or two figures cannot therefore contain more than one figure, that is, units.* Should the number be an imperfect square, its root will be found between two whole numbers, which differ from each other by 1, and whose squares are the next greatest and smallest squares, between which the given number is comprised. Thus, 46 is comprised between 36 and 49, which are the squares of 6 and 7 ; the square root of 46 is then between 6 and 7. Again, 96 is between 81 and 100, the square roots of which are 9 and 10 ; the square root of 96 is between 9 and 10.

Q. In what does the extraction of the square root consist ? How many perfect squares are there which contain only one or two figures ? What are their roots ? How many figures will the square root of a number comprised of one or two figures contain ? Between what numbers will the square root of a number which is not a perfect square be found ? Between what numbers will the square root of 96 be found ? $\sqrt{45}$? $\sqrt{37}$? $\sqrt{24}$?

221. Every number may be considered as composed of a certain number of tens and a certain number of units. Thus, $16 = 1$ ten $+ 6$ units, or $10 + 6$, and $98 = 9$ tens $+ 8$ units $= 90 + 8$.

Now, the square of a number such as 16, will be the product of $10 + 6$ by itself. To see fully of what parts the square will be composed, we will keep the units, tens and *hundreds* of the square separate, as follows :

Commencing on the right, 6 units of the multiplicand, multiplied by 6 units of the multiplier, give 36 units $= 6^2 = $ *square of the units.* 1 ten or 10 of the multiplicand, by 6 units of the multiplier, give 6 tens or $6 \times 10 = $ *product of these tens by the units.* Multiplying now

OPERATION.

$$10+6$$
$$10+6$$
$$\overline{}$$
$$10 \times 6 + 6^2$$
$$10^2 + 10 \times 6$$
$$\overline{10^2 + 2(10 \times 6) + 6^2}$$

by the ten of the multiplier, the first product is 6 tens, or $6 \times 10 = $ *product of the tens by the units;* the next product is $10 \times 10 = 10^2 = 100 = $ *square of the tens.*

As the same result will be obtained for any other number, we conclude, *that the square of a number composed of tens and units contains the square of the tens, plus twice the product of the tens by the units, plus the square of the units.*

Q. Of what parts may every number be considered as composed? What does 16 contain? 25? 59? 125? What does the square of a number composed of tens and units contain?

222. The squares of the numbers

10, 20, 30, 40, 50, 60, 70, 80, 90, 100,

are

100, 400, 900, 1600, 2500, 3600, 4900, 6400, 8100, 10000.

100 is the smallest number composed of three figures, and its square root contains two figures, and is 10; 9999 is the largest number composed of four figures, and its square root is less than 100, being comprised between 99 and 100. Hence, *when the given square contains three or four figures, the root will contain two figures; that is, tens and units.*

Q. What is the square of 10? 20? 40? 60? 90? 100? What is the smallest number composed of three figures? What is the square root of 100? What is the largest number composed of four figures? What is its square root? If a number contain three figures, how many figures will its square root contain? If four figures, how many figures in the root?

EXAMPLES.

1. Extract the square root of 4096.

Since the number contains four figures, its square root will contain two figures; that is, tens and units. But the square of 1 ten is 100; therefore, the square of the tens of the root cannot be found in the two right-hand figures of the given num-

$$4096 \, (\, 64$$
$$36$$

$$6 \times 2 = 124 \,) \;\; 49.6$$
$$496$$
$$\cdots$$

ber. We separate these figures by a dot. Now 4096 is contained between the two squares 3600 and 4900. Its root must then be between 60 and 70. The required root therefore contains 6 *tens*. We place 6 on the right of the given number, from which we separate it by a line. We then subtract its square 36 hundred from 40 hundred, which leaves a remainder 4, to which we bring down the next two figures 96. After subtracting the square of the tens from the given number, the remainder 496 must contain *twice the product of the tens by the units, plus the square of the units*, (Art. 221). But since tens multiplied by units cannot give a less product than *ten*, twice the product of the tens by the units cannot be contained in the right-hand figure 6 of the remainder, which is units simply. This product must therefore be found in the two left-hand figures 49. Dividing this number by twice the tens, we get 4 for the units of the required root. Placing 4 in the units' place of the root, and also on the right of the divisor 12, and multiplying 124 by 4, the product 496 is exactly equal to the remainder. The root sought is therefore 64, which may be proved by multiplying 64 by itself, $64 \times 64 = 4096$.

It will be observed that in multiplying 124 by 4, we first multiply 4 by 4, and obtain 16, *which is the square of the units of the root*, and then 12, which is *twice the tens* of the root, multiplied by 4, the units, gives the double product of the tens by the units. The product 496 is therefore composed of the same parts as the remainder after the first operation.

2. Extract the square root of 1444.

OPERATION.

$$1\overset{.}{4}44\ (38$$
$$9$$

$$3 \times 2 = 68\ )\ 54.4$$
$$544$$

$$\ldots$$

$$8 \text{ units} \times 8 \text{ units} = \text{square of units} = \quad 64$$
Twice tens $\times$ units $= 6 \text{ tens} \overset{.}{\times} 8 \text{ units} = 48 \text{ tens} = 480$

$$544$$

We separate the two right-hand figures as before. There being again 4 figures in the power, the root contains tens and units. The greatest square in 14 is 9, the root of which is 3. The tens of the root will be 3. Subtracting its square 9 from 14, and bringing down the two other figures 44, we have for a remainder 544. Dividing now the two left-hand figures 54 by 6, which is twice the tens of the root, we find it goes 9 times; but 9 is too large, since 54 contains the tens arising from the square of the units, as well as those produced by the double product of the tens by the units. Trying 8 as the next lower number, and placing 8 on the right of 6, and multiplying 68 by 8, the result 544 is exactly equal to the remainder before found. 38 is therefore the root sought.

By examining the operation of multiplying 68 by 8, we see that the square of the units is 64, which contains 6 *tens*, and these tens are carried into the tens resulting from the multiplication of the double tens by the units.

3. Extract the square root of 729.

The given number containing 3 figures, the root will contain tens and units. Separating the two right-hand figures, the greatest square in 7 is 4, whose root is 2. Subtracting the square of 2 from 7, and bringing down 29, the remainder is 329. Dividing 32 by $2 \times 2 = 4$, *the quotient is* 8;

OPERATION.

$$7\overset{.}{2}9\ (\ 27$$
$$4$$

$$2 \times 2 = 47\ )\ 32.9$$
$$329$$

$$\ldots$$

K

but the tens produced by multiplying 8 by itself, ma
result too large. We find 7 to be the units of th
Multiplying 47 by 9, the result is 329. 27 is tl
sought.

4. Extract the square root of 56169.

Since the given number
contains more than four
figures, the square root will
contain more than two. Regarding the numbers as still
composed of tens and units,
the number 56169 will contain 5616 tens and 9 units.
Now, the square of the tens
cannot be comprised in the
two right-hand figures; we

OPERATION.

$$\dot{5}\dot{6}16\dot{9}$$
$$4$$

$$2\times2=43\,)\,\overline{16.1}$$
$$129$$

$$23\dot{\times}2=467\,)\,\overline{326.}$$
$$326\dot{9}$$

$$\cdots\cdots$$

point them off as before, and the square of the ten:
then be found in the remaining figures. But since the
more than two figures left, the tens of the root will ‹
more than one figure, and we may separate the two
hand figures of 561 as if its root alone had to be exti
and in like manner we should continue to divide off th
number into *periods* of two figures from the right, if tl
left contained more than one figure. Since 4 is the g
square contained in 5, the root of which is 2, the firsl
in the root will be 2. Subtracting the square of 2 f
and bringing down the next *period* 61, we have 16:
remainder. Dividing 16 by twice the first figure of th
that is, by 4, we get 3 as the second figure of the *tens*
required root. Placing 3 in the root, and also on th
of the divisor, and multiplying 43 by 3, and subtracti
product as before, we have 32 left. Bringing down th
period 69, and dividing 326 by twice the tens of th
that is, by 23 × 2, we find 7 for the units of the root.
ceeding as before, we find 237 to be the root sough
taining 23 tens and 7 units.

It will be observed, that there will be always as
figures in the root as there are periods.

From the above examples we may deduce the follo

RULE.

I. *Point off the given number into periods of two figures each, commencing on the right. Find the greatest square in the period on the left, and place its root on the right of the given number as the first figure of the root.*

II. *Subtract the square of the root from the number comprising the first period, and to the remainder bring down the second period.*

III. *Double the root found, and place the product on the left of the remainder as a divisor. Divide the remainder, omitting the right-hand figure, by this divisor: the quotient will be the second figure of the root.*

IV. *Place it in the root, and also on the right of the divisor. Multiply the entire divisor by the second figure of the root, and subtract the product from the dividend. To the remainder bring down the next period.*

V. *Double the whole of the root found for a new divisor, and divide as before: the quotient will be the third figure of the root.*

VI. *Continue this operation until all the periods are brought down. Should there be a remainder after the last period has been divided, add ciphers, regarding the terms of the root thus found as decimals.*

Q. How do you extract the square root of a number ? If the number contain more than two figures, but less than five, how many figures will there be in the root ? Will the square of the tens be found in the two right-hand figures ? Why not ? What do you do with these two figures ? If the remaining figures are more than two, what is done ? With which period do you commence to extract the root ? What will the root of this period be ? Where placed ? What is done next ? When the square of the tens is taken from the given number, what will the remainder contain ? What do you annex to the remainder ? What do you do with this remainder ? Why do you divide by twice the tens of the root ? Where do you place the quotient ? Why do you not divide the right-hand figure of the remainder ? Is not the quotient figure too large sometimes ? Why ? What is done ? Why do you place the quotient figure in the root and also in the divisor ? How long is this operation continued ? How will the number of figures in the root correspond with the number of periods ? Should there be a remainder after all the periods are divided, what do you do ? What will the figures of the root be ? Repeat the rule for extracting the square root of whole numbers.

1. Extract the square root of 7569.   Ans. 87.

2. Extract the square root of 76807696.

Ans. 8764 + .

3. Extract the square root of 87567.   Ans. 295917 + .

4. Extract the square root of 170597631.

Ans. 13061 + .

5. Extract the square root of 94307827.

Ans.

6. Extract the square root of 100400730.

Ans.

7. Extract the square root of 5320717.

Ans.

8. Extract the square root of 8000910032.

Ans.

223. Since the square of a vulgar fraction is obtained by multiplying its numerator by itself for a new numerator, and its denominator by itself for a new denominator, it follows, that *to extract the square root of a fraction whose terms are perfect powers, we have only to extract the root of the numerator and also of the denominator.*

Thus, $\sqrt{\frac{9}{16}} = \frac{3}{4}$, since the square root of 9 is 3, and that of 4 is 2.

1. Extract the square root of $\frac{25}{36}$.   Ans. $\frac{5}{6}$.

2. Extract the square root of $\frac{576}{625}$.   Ans. $\frac{24}{25}$.

3. Extract the square root of $\frac{12996}{9025}$.   Ans.

Q. How is the square root of a vulgar fraction extracted? Why is this?

224. Should the numerator of the fraction be an imperfect power, we extract its square root approximately, and *place the* result over the root of the denominator.

Thus, to extract the square root of $\frac{2}{9}$.

OPERATION.

The square root of 2 is 1. Annexing ciphers, and continuing the operation to two places of decimals in the root, we find the result to be $\frac{1\cdot41}{3}+$. By annexing more ciphers, we might carry the operation to any degree of approximation.

$$2.0000\ (\ 1.41+$$
$$1$$
$$\overline{}$$
$$1\times2=24\ )\ 10.0$$
$$96$$
$$\overline{}$$
$$14\times2=281\ )\quad 40.0$$
$$281$$
$$\overline{}$$
$$19$$
$$\overline{}$$
$$\sqrt{\tfrac{2}{3}}=\tfrac{1\cdot41}{3}+.$$

Q. How is the square root extracted when the numerator is an imperfect square? How far may the operation be extended?

EXAMPLES.

1. Extract the square root of $\frac{88}{50}$.   Ans.
2. Extract the square root of $2\frac{5}{6}$.   Ans.
3. Extract the square root of $\frac{5}{4}$ of $7\frac{4}{25}$. Ans.

225. Should the denominator of the fraction be an imperfect power, we multiply the two terms of the fraction by the denominator. By this operation, the denominator is made a perfect square, and the root of the fraction may be extracted as in the last article.

Thus, extract the square root of $\frac{3}{5}$.

Multiplying numerator and denominator by 5, the denominator becomes 25, the square root of which is 5. Extracting the square root of 15 to three places of decimals, it becomes 3.872. The root sought is $\frac{3\cdot872}{5}$.

OPERATION.

$$\frac{3}{5}=\frac{3\times5}{5\times5}=\frac{15}{25}$$
$$\sqrt{15}=3.872$$
$$\sqrt{25}=\ 5$$
$$\sqrt{\tfrac{15}{25}}=\frac{3\cdot872}{5}.$$

Q. How is the square root of a fraction extracted when its denominator is an imperfect power? Is the value of the fraction changed by this operation? Why do you multiply by the denominator?

EXAMPLES.

1. Extract the square root of $\frac{1}{4}$.   Ans.
2. Extract the square root of $3\frac{1}{3}$.   Ans.
3. Extract the square root of $\frac{4}{9}$.   Ans.

226. Since a decimal may be expressed as a vulgar fraction by placing 1 as its denominator, with as many ciphers annexed as there are decimal places, we may extract the square root of a decimal by the principles just explained.

Should the given decimal contain an uneven number of places, a cipher must be annexed to make them even, and by this means the denominator will be made a perfect square. The square root of the decimal may then be extracted as if it were a whole number, care being taken to point off as many decimal places in the root as there are periods in the given decimal.

Thus, extract the square root of .5.

The number of decimal places being odd, we annex a cipher, making the decimal $.50 = \frac{50}{100}$. Extracting the root of 50, it is .7 $= \frac{7}{10}$. By annexing two more ciphers, the decimal becomes $.5000 = \frac{5000}{10000}$, the root of which is $.70 = \frac{70}{100}$. By continuing this operation we may approximate as near as we please to the root of $\frac{1}{2}$.

OPERATION.

$$.50\ (\ .707+$$
$$49$$
$$\overline{\hspace{1cm}}$$
$$.1407\ )\ 10000$$
$$9849$$
$$\overline{\hspace{1cm}}$$
$$151$$

Q. How may the square root of a decimal fraction be extracted?

EXAMPLES.

1. Extract the square root of .25.   Ans. .5.
2. Extract the square root of .42573.   Ans.
3. Extract the square root of .0024763.   Ans.
4. Extract the square root of 21.935.   Ans. 4.683.
5. Extract the square root of .542.   Ans. .736.
6. Extract the square root of .0054.   Ans. .073.

Note.—By reducing the vulgar fractions in articles 224 and 225 to, equivalent decimals, their square root might be extracted as has just been explained for decimal fractions.

EXTRACTION OF THE CUBE ROOT.

227. THE extraction of the cube root of a number consists in finding a number which, multiplied by itself twice, will produce the given number.

Thus, the cube root of 27 is 3, since $3 \times 3 \times 3 = 27$.

We have seen (Art. 219) that there are only nine perfect cubes expressed by one, two, or three figures, viz :

1, 8, 27, 64, 125, 216, 343, 512, 729,

the cube roots of which are the first nine numbers,

1, 2, 3, 4, 5, 6, 7, 8, 9.

Hence, *the cube root of every number composed of less than four figures cannot contain more than one figure, that is, units.*

The root of an imperfect cube will be found between two numbers which differ from each other by one, and whose cubes are the next greatest and smallest cubes, between which the given number is comprised. Thus, 130 is comprised between 125 and 216, the cube roots of which are 5 and 6. The cube root of 130 will therefore be between 5 and 6.

Q. In what does the extraction of the cube root of a number consist ? What is the cube root of 27 ? Why 3 ? How many perfect cubes may be expressed by one, two, or three figures ? What are their roots ? How many figures will the cube root of a number composed of fewer than four figures contain ? What will the root be ? Between what numbers will the root of an imperfect cube be found ?

228. To explain the method of extracting the cube root of a number, let us see of what parts the cube is composed

The cube of the number 16 is the product of 16 or its equal $10+6$ by itself twice. Performing the operation as in Art. 121, we find the result as follows:

OPERATION.

$$10+6$$
$$10+6$$

$$10\times 6+6^2$$
$$10^2+10\times 6$$

$$10^2+2(10\times 6)+6^2=\text{square of } (10+6)$$
$$10+6$$

$$10^2\times 6 +2(10\times 6^2)+6^3$$
$$10^3+2(10^2\times 6)+ \quad 10\times 6^2$$

$$10^3+3(10^2\times 6)+3(10\times 6^2)+6^3=\text{cube of } (10+6).$$

The first multiplication gives the square of $10 + 6$, as found in Art. 121. Multiplying again by $10+6$, we obtain the cube of $10+6$.

But upon examining this result we find that it is composed as follows:

1st, $10^3=$cube of the tens.

2d, $3(10^2\times 6)=$three times the square of the tens multiplied by the units.

3d, $3(10\times 6^2)=$three times the square of the units multiplied by the tens.

4th, $6^3=$cube of the units.

And as the same may be proved for every other such number, we conclude, *that the cube of a number composed of tens and units contains the cube of the tens, plus three times the square of the tens multiplied by the units, plus three times the square of the units multiplied by the tens, plus the cube of the units.*

Thus, the cube of 25 is as follows:

PROOF.

25
25
———
125
50
———
625
25
———
3125
1250
———
15625

OPERATION.

$(2 \text{ tens})^3 = 20^3 = 8000$
$3(20)^2 \times 5 \quad = 6000$
$3(20) \times 5^2 \quad = 1500$
$5^3 \qquad = \ 125$
———
15625

Q. Of what is the cube of a number comprising tens and units composed? How will you find the cube of 25? Of 15? Of 75? Of 125?

EXAMPLES.

1. Extract the cube root of 15625.

Since the given number contains more than three figures, its root will contain more than one, and will be composed of tens and units. Again, since the cube of 10 is 1000, the cube of the tens

OPERATION.

$15625 \ (\ 25$
$\quad 8$
——————
$2^2 \times 3 = 12 \ ) \ 76.25$
$\qquad 25^3 = 15625.$

of the root cannot be found in the three right-hand figures. We point them off as in the square root. Now the greatest cube in 15 is 8, the root of which is 2. Subtracting 8 from 15, and bringing down the next period, the remainder 7625 contains 3 times the square of the tens multiplied by the units, plus 3 times the square of the units multiplied by the tens, plus the cube of the units. Cutting off two figures on the right, the 76 which remains contains 3 times the square of the tens multiplied by the units; and if it be divided by 3 times the square of the tens of the root, that is, by 12, the quotient 5 will be the units of the root. 12 goes in 76, 6 times; but 6 being too large, we try 5, and find 25 to be the cube root of the given number, since $25^3 = 15625$.

2. Extract the cube root of 596947688.

K 2

We commence by pointing off the given number into periods of 3 figures each, from the right. 512 is the greatest cube in the first period, the root of which is 8. Subtracting 512 from 596, and bringing down the next period, we cut off the two

$$596947688\,(\,842$$
$$512$$
$$\overline{}$$
$$8^2 \times 3 = 192\,)\ 849{,}47$$
$$596947$$
$$84^3 = 592704$$
$$\overline{}$$
$$84^2 \times 3 = 21168\,)\ 42436{,}88$$
$$842^3 = 596947688.$$

right-hand figures, and divide 849 by $192 = 8^2 \times 3$. This gives 4 for the second figure of the root. Cubing 84 and subtracting the cube from the two first periods, we have 4243 for a remainder. Bringing down the next period, and cutting off 88, we divide by 3 times the square of the root found, that is, by 3×84^2, and the quotient gives the units of the root. If the result be correct, and the number be a perfect cube, 842^3 will be equal to the given number.

From these examples we deduce the following

RULE.

I. *Point off the given number into periods of three figures each, commencing on the right.*

II. *Find the greatest cube contained in the left-hand period. Its cube root will be the first figure of the root sought. Subtract its cube from the first period, and bring down the next period to the remainder, cutting off the two right-hand figures.*

III. *Divide the figures not cut off by three times the square of the root found: the quotient will be the second figure of the root. Cube the whole root found, and subtract it from the first two periods of the given number. If the cube be larger than the number in these periods, the last figure in the root must be diminished; but if less, bring down the next period, cutting off the two right-hand figures as before.*

IV. *Divide this last remainder by three times the cube*

of the whole root found: the quotient (if not too large) will be the next figure in the root. Continue in this way until all the periods are brought down. Then cube the whole root, and subtract it from the given number. If it be a perfect cube, there will be no remainder.

Q. How do you extract the cube root of a number? Why point off three figures? How many figures does the cube of 10 contain? If the number contains more than three figures, will the root contain one or more figures? Where do you commence to extract the root? When the cube of the first figure of the root is subtracted from the first period, what does the remainder contain? By what do you divide? Why? What next? If the given number be a perfect cube, will there be any remainder when the cube of the root is taken from it? Why?

EXAMPLES.

1. Extract the cube root of 13824.   Ans. 24.
2. Extract the cube root of 5832.   Ans. 18.
3. Extract the cube root of 1157625.   Ans. 105.
4. Extract the cube root of 500763009.   Ans.
5. Extract the cube root of 97605732189.   Ans.
6. Extract the cube root of 2030421.   Ans.

228. Since the cube of a vulgar fraction is formed by cubing its numerator and denominator, it follows, *that to extract the cube root of a vulgar fraction, we extract the cube root of its numerator, and then of its denominator.*

Thus, the cube root of $\frac{27}{64} = \frac{3}{4}$, since $3 = \sqrt[3]{27}$ and $4 = \sqrt[3]{64}$. If the numerator be not a perfect cube, we extract its nearest root, and place the result over the cube root of the denominator.

Thus, the cube root of $\frac{143}{343}$ is $\frac{5.22}{7}$, since 5.22 is the approximate root of 143, and 7 is the cube root of 343.

Again, if the denominator be an imperfect power, we multiply both terms of the fraction by the square of the denominator. By this operation the denominator is made a perfect cube; we may then extract the root of both terms as has just been explained.

Thus, to extract the cube root of $\frac{3}{7}$. Multiplying numerator and denominator by the square of 7, that is, 49, the fraction becomes $\frac{147}{343} = \frac{3}{7}$, the cube root of which is $\frac{5.27}{7}$,

Q. How is the cube of a vulgar fraction obtained ? How is the cube root of a vulgar fraction found ? If the numerator be an imperfect cube ? If the denominator be an imperfect cube ? Why do you multiply by the square of the denominator ? Does this operation change the value of the fraction ? Why not ?

EXAMPLES.

1. Extract the cube root of $\frac{125}{729}$.   Ans. $\frac{5}{9}$.
2. Extract the cube root of $3\frac{4}{9}$.   Ans.
3. Extract the cube root of $5\frac{3}{64}$.   Ans.
4. Extract the cube root of $\frac{129}{42}$.   Ans.
5. Extract the cube root of $\frac{1}{2}$.   Ans.
6. Extract the cube root of $3\frac{1}{2}$.   Ans.
7. Extract the cube root of $\frac{5}{11}$.   Ans.
8. Extract the cube root of $\frac{10}{9}$.   Ans.

230. To extract the cube root of a decimal, we must, if necessary, annex ciphers, so that the number of decimal places shall be 3, 6, 9, &c. We then proceed as in whole numbers, pointing off as many decimal places in the root as there are periods in the given decimal.

EXAMPLES.

1. Extract the cube root of 6.54.   Ans. 1.87.
2. Extract the cube root of .0006.   Ans. .08+.
3. Extract the cube root of .5.   Ans.
4. Extract the cube root of 1.23.   Ans. 1.07+.
5. Extract the cube root of 24.17.   Ans.
6. Extract the cube root of 241.7.   Ans.
7. Extract the cube root of 2.417.   Ans.

Q. How is the cube root of a decimal extracted ? How many decimal places will there be in the root ?

APPLICATIONS UPON THE FOREGOING RULES.

1. The intensity of the same light varies for different distances in the inverse ratio of the squares of those distances. Supposing the intensity of a light at 10 feet distance to be *1, what numbers will represent* its intensity at 200, 300, *400, 500 feet ?*   Ans. $\frac{1}{400}, \frac{1}{900}, \frac{1}{1600}, \frac{1}{2500}.$

2. The intensity of gravity follows the same law as just announced. What will its intensity be equal to at 2, 3, 4, 5 feet distance, that at the unit's distance being represented by 1 ? Ans. $\frac{1}{4}$, $\frac{1}{9}$, $\frac{1}{16}$, $\frac{1}{25}$.

3. What is the weight of a block of common stone which measures 10 ft. 2′ in length, 4 ft. 5′ broad, and 2 ft. 3′ thick, the weight of a cubic foot being 2520 ounces avoirdupois ?
Ans. 15907 + lbs.

4. What is the weight of a bar of iron measuring 18½ feet long, 4′ thick, 11½ inches broad, the weight of a cubic foot being 7645 ounces ? Ans.

5. What is the weight of a load of clay, which measures 12 ft. 3′ long, 4 ft. 2′ broad, and 2½ feet deep, the weight of a cubic foot being 2160 ounces ?
Ans.

6. The following table exhibits the area in square miles of the different countries in the world, and the States of the American Union : find the side of the squares whose areas shall be equivalent to the given areas.

Square Miles.

England, ... 51,500 .... Ans. 226.9 miles.
Scotland, ... 30,000 .... Ans. 173.1.
Wales, ... 8,500 .... Ans. 92.1.
Ireland, ... 31,000 .... Ans. 176 + .
North America, .. 7,950,000 .... Ans. 2819.3.
South America, .. 7,050,000 .... Ans.
Europe, ... 3,500,000 .... Ans.
Asia, ... 16,000,000 .... Ans.
Africa, ... 11,000,000 .... Ans. 3316.6 + .
France, ... 205,000 .... Ans.
Spain, ... 183,000 .... Ans.
Maine, ... 35,000 .... Ans.
New Hampshire, ... 9,491 .... Ans.
Vermont, ... 8,000 .... Ans.
Massachusetts, ... 7,800 .... Ans.
Rhode Island, ... 1,225 .... Ans. 35.
Connecticut, ... 4,764 .... Ans.
New York, ... 47,000 .... Ans.
New Jersey, ... 8,320 .... Ans.

Square Miles.

Pennsylvania,	 46,000	....	Ans.
Delaware,	 2,100	....	Ans. 45.82.
Maryland,	 9,356	....	Ans.
Virginia,	 70,000	....	Ans. 264.5+.
North Carolina,	 50,000	....	Ans.
South Carolina,	 33,000	....	Ans.
Georgia,	 62,000	....	Ans. 249 *nearly.*
Alabama,	 51,770	....	Ans.
Mississippi,	 48,000	....	Ans.
Louisiana,	 48,320	....	Ans.
Ohio,	 44,000	....	Ans. 209.7+.
Kentucky,	 40,000	....	Ans. 200.
Tennessee,	 45,000	....	Ans.
Michigan,	 60,000	....	Ans. 244.9+.
Indiana,	 36,400	....	Ans.
Illinois,	 55,000	....	Ans.
Missouri,	 64,000	....	Ans. 253.3+.
Arkansas,	 55,000	....	Ans.

Q. Which country in the world has the largest extent? Which State in the American Union has the largest area? Which the smallest? How does the extent of England compare with Virginia?

ARITHMETICAL PROGRESSION.

231. AN ARITHMETICAL PROGRESSION *is a series of numbers, increasing or decreasing by a common difference.*

Thus, the series

2, 5, 8, 11, 14, 17, 20, 23, 26, 29, &c.,

in which each number is formed from the preceding, by adding the *common difference 3, is an increasing arithmetical series* or *progression;* and the series

29, 26, 23, 20, 17, 14, 11, 8, 5, 2,

in which each number is formed from the preceding by subtracting the common difference 3, is a *decreasing arithmetical series* or *progression.* The numbers which form a series or progression are called its *terms.* The first and last terms are the *extremes.* The intermediate terms are called the means.

We may readily form an increasing arithmetical series when we know the first term and the common difference; for, we have only to add the common difference to the first term to find the second; then to the second to find the third; to the third to find the fourth, and so on.

Thus, if the first term be 1, and the common difference 2, we shall have

$$1, 3, 5, 7, 9, 11, 13, 15, 17, 19, \&c.,$$

which is an increasing arithmetical series. To find the terms of a decreasing arithmetical series, when we know the first term and the common difference, we subtract the common difference from the first term to find the second, then from the second to find the third, and so on. Thus,

$$21, 19, 17, 15, 13, 11, 9, 7, 5, 3, 1,$$

is a decreasing arithmetical series, the common difference being 2.

Q. What is an arithmetical progression? Give an example of such a progression. What is an increasing arithmetical series? How is each term formed in it? What is a decreasing arithmetical series? How is it formed? What are the numbers called which compose a progression? What are the first and last terms called? Which are the means? How may you form an increasing arithmetical series when you know the first term and the common difference? Form such a series when the first term is 3 and the common difference 4. How is a decreasing arithmetical series formed? Form such a series when the first term is 20 and the common difference 5.

232. We have seen (Art. 231), that in an increasing arithmetical series the second term is formed from the first by adding to the first the common difference. Thus, if the first term be 2, and the common difference 3, the second term will be $2+3=5$.

The third term is formed by adding the common difference to the second term, or *twice* the common difference to the first term. The third term will therefore be $2+3\times 2 =8$.

The fourth term is found by adding 3 times the common difference to the first term; that is, fourth term $=2+3\times3$ $=11$.

The fifth term $=2+3\times4=14$. And so on for any term.

Hence, we conclude that, in general, *any term of an arithmetical progression may be found by adding to the first term the common difference taken as many times less 1 as there are terms in the series.*

EXAMPLES.

1. Find the twentieth term of an arithmetical progression whose first term is 1, and common difference 5.

There being twenty terms, 20 less 1 will be 19, which, multiplied into the common difference 5, and added to the first term 1, the twentieth term is found to be 96.

OPERATION.

$1+5\times19=96$ Ans.

2. Find the fifteenth term of an arithmetical progression, whose first term is 0, and common difference 4.

Since the first term is 0, the term sought will be as many times the common difference as there are terms less 1.

OPERATION.

$0+4\times14=4\times14=56$.

3. Find the hundredth term of the progression, 1, 4, 7, 10, 13, 16, &c.   Ans. 298.

4. Find the forty-ninth term of the progression 4, 9, 14, 19, 24, &c.   Ans. 244.

Q. How may any term of an increasing arithmetical series be found when you know the first term and the common difference? What is the third equal to, when the first term is 2 and common difference 5? Fourth term? Fifth term?

233. If the first and last terms of an arithmetical progression be known as well as the number of terms, we may readily find the common difference; for, since the last term is equal to the first term added to the common difference taken as many times less 1 as there are terms, it follows, *that the common difference will be equal to the difference of the extremes divided by the number of terms less 1.*

EXAMPLES.

1. Find the common difference in the arithmetical series whose first term is 3, last term 19, and number of terms nine.

Subtracting the first term 3, from the last term 19, and dividing by 8, which is 1 less than the number of terms, the common difference is found to be 2.

OPERATION.

$$\frac{19-3}{8}=\frac{16}{8}=2=\text{common difference.}$$

PROOF.

3, 5, 7, 9, 11, 13, 15, 17, 19.

2. Find the common difference in arithmetical series, the extremes being 10 and 70, and the number of terms twenty-one. Ans. 3.

3. Find the common difference in the arithmetical progression whose first term is 4, last term 11, and number of terms ten. Ans. $\frac{7}{9}$.

Q. How may you find the common difference when you know the first and last terms, and the number of terms of an arithmetical progression?

234. It follows from what has just been explained (Art. 233), that when two numbers are given, we may insert between them as *extremes*, any number of *means* of an arithmetical progression; for, knowing the first and last terms, and the number of terms to be inserted, we can find the common difference, and the several means may then be formed by adding the common difference as in Art. 232.

Thus, if we wished to insert between 2 and 14, five arithmetical means, the series would contain seven terms; hence, the common difference would be $\frac{14-2}{6}=2$ (Art. 233). The series would then be 2, 4, 6, 8, 10, 12, 14, in which there have been five terms inserted between the first and last terms. We see that the divisor 6 is one more than the number of terms to be inserted. Hence, *to insert any number of arithmetical means between two given extremes, divide the difference of the extremes by the number of means to be inserted, plus 1, for the common difference, and then form the series as before explained.*

EXAMPLES.

1. Insert eight arithmetical means between 4 and 11.

Ans. Com. dif. $\frac{7}{9}$.

means, $4\frac{7}{9}$, $5\frac{5}{9}$, $6\frac{3}{9}$, $7\frac{1}{9}$, $7\frac{8}{9}$, $8\frac{6}{9}$, $9\frac{4}{9}$, $10\frac{2}{9}$.

2. Insert fifteen arithmetical means between 5 and 100.

Ans.

Q. How may you insert any number of arithmetical means between two given extremes ?   How do you find the common difference ?

235. A fundamental property of an arithmetical progression is, that *the sum of the extremes is always equal to the sum of any two terms taken equi-distant from the extremes.*

This will appear by arranging the terms of any series in an inverse order.   Thus,

1	3	5	7	9	11	13	15
15	13	11	9	7	5	3	1
16	16	16	16	16	16	16	16

The sum of the extremes 1 and 15 is 16, and the same result is found for every pair of equi-distant terms.

If the number of terms in the series be odd, the sum of the extremes will be equal to twice the middle term.   Thus,

1	3	5	7	9
9	7	5	3	1
10	10	10	10	10

It follows from the above property, that the sum of the two series will be equal to the sum of the extremes, 16, taken as many times as there are terms, that is, eight times; and hence, *in an arithmetical progression, the sum of its terms is equal to half the sum of the extremes, multiplied by the number of terms.*

EXAMPLES.

1. Find the sum of the terms of an arithmetical series whose first term is 3, last term 19, and number of terms nine.

$\frac{2+19}{2} \times 9 = \frac{22}{2} \times 9 = 11 \times 9 = 99 =$ sum of the series.

$\frac{2+19}{2}$ will be the half sum of the series, which, being multiplied by 9, the number of terms, gives 99 for the sum of the series.

2. Find the sum of an arithmetical progression, whose first term is 2, last term 100, and number of terms fifty.
Ans. 2550.

3. Find the sum of an arithmetical progression, whose first term is 10, common difference 3, and number of terms twenty-one.
Ans. 840.

Q. In an arithmetical progression composed of an even number of terms, what is the sum of the extremes equal to? If the number of terms be odd? What is the sum of an arithmetical series equal to?

GEOMETRICAL PROGRESSION.

236. A GEOMETRICAL PROGRESSION *is a series of numbers increasing or decreasing by a constant ratio.*

Thus, the series

1, 2, 4, 8, 16, 32, 64, &c.

in which each term is formed from the preceding, by *multiplying* by the number 2, *is an increasing geometrical progression;* and the series

64, 32, 16, 8, 4, 2, 1,

in which each term is formed from the preceding by *dividing* by the number 2, *is a decreasing geometrical progression.*

The number by which we continually multiply or divide is called the *common ratio.*

2, 6, 18, 54, 162, &c.

is an *increasing* geometrical progression.

162, 54, 18, 6, 2,

is a *decreasing* geometrical progression, the common ratio being 3.

Q. What is a geometrical progression? What is an increasing geometrical progression? Decreasing? Give an example of an increasing series. Of a decreasing series. What is the common ratio? In the series 2, 6, 18, &c., what is the common ratio? In the series 54, 18, 6, 2, what is the common ratio? In the series 5, 20, 80, &c.? In the series 100, 20, 4?

237. From the manner in which an increasing geometrical series is formed, we see that

2d term = 1st term multiplied by the common ratio.

3d term = 2d term multiplied by the common ratio, or the 1st term multiplied by the *square* of the common ratio.

4th term = 3d term multiplied by the common ratio, or the 1st term multiplied by the 3d power of the common ratio.

5th term, in like manner = the 1st term multiplied by the 4th power of the common ratio, and so with the other terms.

Hence, we conclude, *that any term of an increasing geometrical progression is equal to the first term multiplied by the common ratio raised to a power one less than the number of terms.*

NOTE. — In a decreasing series, we would divide by the common ratio raised to a power less 1 than the number of terms.

EXAMPLES.

1. Find the sixth term of the progression 1, 3, 9, 27, &c.

1 being the first term, 3 the common ratio, and the number of terms six, the sixth term will

OPERATION.

$1 \times 3^5 = 1 \times 243 = 243.$

be equal to the first term 1, multiplied by the common ratio 3 raised to the fifth power, 5 being 1 less than the number of terms. But $3^5 = 3 \times 3 \times 3 \times 3 \times 3 = 243.$ And $1 \times 243 = 243.$

2. Find the seventh term of the geometrical progression, whose first term is 3 and common ratio 2.

Ans. 192.

3. Find the tenth term of the progression 2, 10, 50, 250, &c.   Ans. 3906250.

238. *In every geometrical progression, the product of the extremes is always equal to the product of any two terms taken equi-distant from the extremes, when the number of terms is even; and to the square of the middle term when the number of terms is odd.*

Thus, arranging the terms of a series in an inverse order,

1	2	4	8	16	32
32	16	8	4	2	1
32	32	32	32	32	32

the product of the extremes 1 and 32 is 32, and the same result is found for the product of every pair of equi-distant terms.

In the following series, which has an odd number of terms, we have

2	6	18	54	162
162	54	18	6	2
324	324	324	324	324

the product of the extremes 2 and 162 is equal to the square of the middle term 18 or 18×18.

Q. What is the product of the extremes of a geometrical progression equal to?   If the number of terms be odd?

239. In a geometrical progression, *the sum of the series is equal to the last term multiplied by the ratio; this product diminished by the first term, and the remainder divided by the ratio less one.*

EXAMPLES.

1. What is the sum of the geometrical series whose first term is 2, ratio 3, and number of terms ten?

OPERATION.

10th term $= 2 \times 3^9 = 39366$ (Art. 237).

Sum of the series $= \frac{39366 \times 3 - 2}{3 - 1} = \frac{118096}{2} = 59048$.

We must first find the last term (Art. 236), and knowing the last term, we can find the sum of the series by multiplying it by the ratio 3; then subtracting the first term 2 from the product, and dividing the remainder by the ratio minus 1.

2. The first term of a geometrical series is 1, the ratio 2, and last term 32; what is the sum of the series?

Ans. 63.

3. Find the sum of the first twenty terms of the series 3, 9, 27, 81, &c. Ans.

Q. What is the sum of a geometrical progression equal to?

APPENDIX.

MENSURATION;

OR, THE

APPLICATION OF ARITHMETIC TO THE MEASURE-MENT OF PLANE AND SOLID BODIES.

240. BEFORE we can apply arithmetic to the measurement of geometrical figures, we must make use of some known magnitude which may serve as a *common measure.* Such a common measure is called an *unit.* (Art. 2).

The unit is always of the same kind as the magnitudes which are measured; that is, it must be a *line,* if lines are measured; a plane, if planes are measured; and a solid, if solids are measured.

Thus, to find the length of the line A B, we make use of a line C D of known length, say 1 inch, 1 foot, or 1 yard, and apply it to the line A B. It is contained in it six times. The *number* 6 is then the length of the line A B, and the length will be 6 *inches,* 6 *feet,* or 6 *yards,* according as C D is 1 inch, 1 foot, or 1 yard. C D is the unit for lines, or *linear unit.*

Again, if the figure to be measured were the *square* A B C D, we would use a square, U, the side of which is 1 inch, 1 foot, or 1 yard, &c., as the unit, and apply it to the square A B C D.

(237)

It is contained in it 9 times. The number 9 expresses the *area*, or content of the square A B C D, and the area will be 9 square inches, 9 square feet, 9 square yards, &c., according as the unit U is a square inch, a square foot, a square yard, &c. Each of the sides of the square A B C D con‹ tains 3 *linear* units, and as there are 3 × 3 *square* units in the given square, it is evident the area, which is 9, will be found by multiplying one side of the square by itself.

If the figure to be measured be a solid, we make use of a solid with six equal square faces, called a *cube,* as the *unit;* and the solidity of the body will be expressed in *cubic inches, feet,* or *yards,* according as the sides of the cubic unit are inches, feet, or yards.

Thus, the cube A B contains the cubic unit U, 27 times, and the number 27 expresses the solidity of the larger cube, and will represent 27 cubic inches, feet or yards, according as the unit is a cubic inch, foot or yard. The base of the large cube contains 3 × 3=9 square feet; 9 cubic units can there‹ 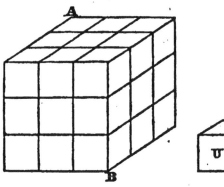 fore be placed on it; and since there are 3 linear units in the side A C, there will be 3 × 3 × 3=27 cubic units in the solid. *Hence the solidity of a body is always equal to the product of its length, breadth and thickness.*

Q. What is necessary before we can apply arithmetic to the measure‹ ment of geometrical figures ? What is this common measure called ? What kind of magnitude must the unit be ? If lines are measured ? Planes ? Solids ? How do you find the length of a line ? What will the number which expresses its length be, if the unit is an inch ? Foot ? Yard ? What is the unit for lines called ? How do you measure a square ? What is meant by an area ? What will express the area of a square ? When will the area be square inches ? Square feet ? Square yards ? How is the area of a square found ? How do you measure a solid body ? What is the unit called ? What will express the solidity of the body ? How is the solidity of a body always determined ? When will the solidity be expressed in cubic inches ? Cubic feet ? Cubic yards, &c. ?

241. A *triangle* is a plane figure bounded by three straight lines.

A B C is a *trian- gle.* The lines A B, B C, and A C, are called *sides.* The spaces A BC, BAC, A C B, formed by two sides respect- ively with each other, are called *angles.* There are three angles in

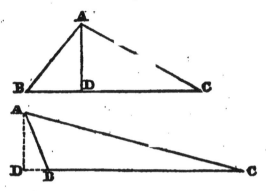

every triangle. The side upon which the triangle stands is its *base.* B C is the base of the triangle A B C. The *alti- tude* of a triangle is the perpendicular let fall from the oppo- site angle upon the base or the base produced. A D, perpen- dicular to B C, is the altitude of the triangle A B C. When one line is perpendicular to another line, the angle formed is a *right* angle. A D C and A D B are *right angles,* and the triangles A B D, A D C, are called *right-angled trian- gles.* The side A B, opposite the right angle A D B, is called the *hypothenuse.* The side A C is the hypothenuse of the right-angled triangle A D C.

The following propositions are of frequent use :

1. *The area of a triangle is equal to its base multiplied by one-half of its altitude.*

2. *In every right-angled triangle, the hypothenuse is equal to the square root of the sum of the squares of the two other sides of the triangle.*

3. *In every right-angled triangle, either side is equal to the square root of the difference of the squares of the hypo- thenuse and other side.*

Q. What is a triangle? What are the angles of a triangle? How many angles in a triangle? What is the base of a triangle? Altitude? What is a right angle? What is a right-angled triangle? What is the hypothenuse? Draw a triangle upon your slate. Draw a right-angled triangle. Which side is the hypothenuse? Which angle is the right angle? Draw the altitude. Which is the base? How is the area of a triangle found? How do you find the length of the hypothenuse of a right-angled triangle? Either side?

L

1. Find the area of a triangle, whose base is 100 feet, and altitude 16 feet.

We multiply the base, which is 100 feet, by 8, which is half the altitude, and the product, 800 *square feet*, is the area sought.

OPERATION.
100 base.
  8 half altitude.
———
Ans. 800 square feet.

2. Find the length of the hypothenuse of a right-angled triangle, the two sides which contain the right angle being, one 20, and the other 30 yards.

We first square 20, and add this square to the square of 30 ; the square root of 1300 = 36.05 + is the length of the hypothenuse sought.

OPERATION.
$20 \times 20 = 400 =$ square of 20.
$30 \times 30 = 900 =$ square of 30.
———
$1300 =$ sum of squares.

Ans. $= \sqrt{1300} = 36.05 +$ yards.

3. Find the height of the upright which supports the ridge-pole of a house which is 40 feet deep, and whose rafters are 22 feet long.

The rafter will be the hypothenuse of a right-angled triangle ; the upright will be one of the sides, and half the depth of the house the other side. Subtracting the square of half the depth from the square of the length of the rafter, we get

OPERATION.
$22 \times 22 = 484$
$20 \times 20 = 400$
———
  84

Ans. $= \sqrt{84} = 9.1 +$ feet.

84 feet for the difference of these squares. The square root of this difference, which is 9.1 + feet, is the height of the upright.

4. The base of a triangle measures 14½ feet, and its altitude 10 yards : what is its area? Ans. 217½ square ft.

5. What is the number of square yards in a triangle whose base is 100 feet, and altitude 95 feet?

Ans. 527⅗.

6. Find the length of the rafters of a house which is 100 feet deep, and upright 20 feet.   Ans.

7. Find the height of the upright of a house whose depth is 45 feet, and length of rafter $17\frac{1}{4}$ feet.

Ans.

OF QUADRILATERALS.

242. A *Quadrilateral* is a plane figure bounded by four straight lines. There are five varieties of quadrilaterals.

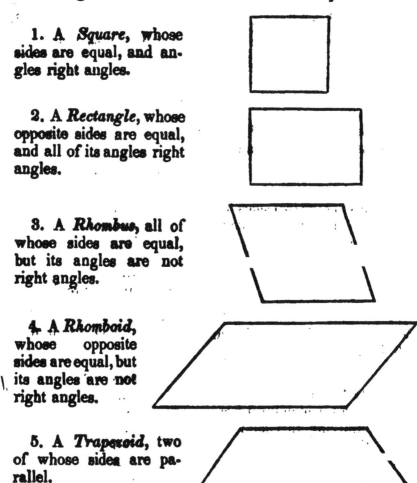

1. A *Square*, whose sides are equal, and angles right angles.

2. A *Rectangle*, whose opposite sides are equal, and all of its angles right angles.

3. A *Rhombus*, all of whose sides are equal, but its angles are not right angles.

4. A *Rhomboid*, whose opposite sides are equal, but its angles are not right angles.

5. A *Trapezoid*, two of whose sides are parallel.

If either side of a square or rectangle be taken for its *base*, the adjacent side is its altitude.

The altitude of a rhomboid or rhombus is the perpendicular distance between the side taken as the base, and the side opposite.

The altitude of a trapezoid is the perpendicular distance between its two parallel sides.

The area of a square, rectangle, rhombus or rhomboid, is found by multiplying its base by its altitude.

The area of a trapezoid is found by multiplying the sum of its parallel sides by half its altitude.

Q. What is a quadrilateral? How many varieties? What is the square? Rectangle? Rhombus? Rhomboid? Trapezoid? What is the altitude of these figures? Draw each of them upon your slate. Point out the base of each. The altitude. How do you find the area of a square? Rectangle? Rhombus? Rhomboid? Trapezoid?

EXAMPLES.

1. What is the area of a rectangle, the adjacent sides of which measure 20 feet and 15 feet?

OPERATION.

Multiplying the two sides together, the product, 300 square feet, is the answer.

$$\begin{array}{r} 20 \\ 15 \\ \hline 100 \\ 20 \\ \hline \end{array}$$

Ans. 300 square feet.

2. What is the area of a trapezoid, whose parallel sides measure respectively 45 feet and 20 feet, and whose altitude is 6 feet?

OPERATION.

Taking the sum of the two parallel sides, which is 65 feet, and multiplying this sum by 3 = ½ the altitude, the result, 195 square feet, is the answer.

$$\begin{array}{r} 45 \\ 20 \\ \hline 65 \\ 3 \\ \hline \end{array}$$

Ans. 195 square feet.

3. What is the area of a rhomboid, whose base is 100 feet, and altitude 5 yards?   Ans. 1500 square ft.

4. What is the area of a rhombus, whose base is 75 feet, and altitude 17 feet?   Ans. 1275 sq. ft.

5. What is the area of a trapezoid, whose parallel sides measure respectively 10 yards and 144 feet, and whose altitude is 70 feet?   Ans. 6090 sq. ft.

OF THE CIRCLE.

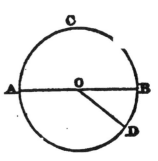

243. A *Circle* is a plane figure, bounded by a curve-line, all of whose points are equally distant from a point within, called the *centre*. The curve A C B D is the *circumference ;* O is the *centre.* The distance O D from the centre to the circumference is called the radius. A line passing through the centre, and terminated on both sides in the circumference, is called the *diameter.* A B is the diameter. The diameter is twice the radius.

The circumference of every circle is found by multiplying its diameter by 3.1416.

The diameter of every circle is found by dividing its circumference by 3.1416.

The area of every circle is found by multiplying the square of its radius by 3.1416.

Q. What is a circle? Draw one upon your slate. Which is the circumference? Diameter? Radius? What is the radius equal to? What is the circumference of a circle equal to? Its diameter? Its area?

EXAMPLES.

1. Find the circumference of a circle whose radius is 10 feet.

Multiply 3.1416 by twice the radius, since the diameter is twice the radius.

OPERATION.

$$\begin{array}{r} 3.1416 \\ 20 \\ \hline \end{array}$$

Ans. 62.8320 feet.

2. Find the diameter of a circle, whose circumference is 1000 feet.

OPERATION.

We divide 1000 feet by 3.1416, by annexing four ciphers. The quotient is 318+feet, which is the length of the diameter sought.

$$3.1416) 1000.0000 (318+\text{feet.}$$
94248

,.57520
31416

261040
251328

..9712

3. Find the area of a circle whose radius is 5 feet.

OPERATION.

We first square the radius, and then multiply by 3.1416.

$5 \times 5 = 25$ square feet.
3.1416

Ans. $= 78.5400$ sq. feet.

4. What is the area of a circle whose circumference is 1000 feet? Ans. 8824.754+sq. yds.

5. Find the area of a circle whose diameter is 200 feet. Ans.

6. Find the area of a circle whose radius is 11 feet. Ans.

OF THE SPHERE.

244. A *Sphere* is a solid round body, all the points of whose surface are equally distant from a point within, called the centre.

The surface of a sphere is found by multiplying the square of its diameter by 3.1416.

The solidity of a sphere is found by multiplying its surface by one-sixth of its diameter.

Q. What is a sphere? How is its surface found? Its solidity?

EXAMPLES.

1. What is the surface of the sphere whose diameter is 6 feet?

The square of the diameter is 36 square feet, which, being multiplied by 3.1416, the result is the surface sought.

OPERATION.

$$3.1416$$
$$36$$
$$\overline{18849}$$
$$94248$$
$$\overline{113.0976} \text{ square feet.}$$

2. What is the solidity of a sphere whose radius is 5 feet?

We first find the surface by multiplying the square of the diameter, which is 10, by 3.1416. Then multiplying this result by $\frac{10}{6}$ or $\frac{1}{6}$ of the diameter, the solidity is 523.6 cubic feet.

OPERATION.

$$3.1416 \times 10^2 = 314.1600 = \text{surface.}$$
$$314.16 \times \tfrac{1}{6}0 = 523.6 \quad = \text{solidity.}$$

3. What is the solidity of the sphere whose radius is 15 feet? Ans.

4. What is the surface of the sphere whose radius is 5 feet? Ans.

OF THE CYLINDER.

245. *The convex surface of a cylinder is found by multiplying the circumference of its base by its altitude.*

The solidity of a cylinder is found by multiplying the area of its base by its altitude.

Q. What is the surface of a cylinder equal to? Its solid content?

EXAMPLES.

1. Find the surface of a cylinder, the circumference of whose base is 10 feet, and altitude 5 feet.

OPERATION.

Multiply 10 by the altitude; the surface in square feet is 50.

$$10$$
$$5$$
$$\overline{}$$
Ans. 50 square feet.

2. Find the solid content of a cylinder, whose altitude is 15 feet, and the radius of whose base is 2 feet.

We first find
the area of the
base by Art. 243,
to be 12.5664

$4 \times 3.1416 = 12.5664 =$ area of base.
$12.5664 \times 15 = 188.496 =$ solidity.

square feet. Multiplying this by the altitude 15 feet, the
result is 188.496 cubic feet.

3. Find the convex surface of a cylinder, whose altitude
is 21 yards, and the radius of whose base is 3 feet.

Ans. 131.9472 sq. yds.

4. Find the solidity of a cylinder, whose altitude is 32
feet, and the diameter of whose base is 2 yards.

Ans. 33.5104 cub. yds.

5. Find the solidity of a cylinder, whose altitude is 10
yards, and the circumference of whose base is 1247 feet.

Ans.

6. Find the convex surface of a cylinder, whose altitude
is 12 feet, and the circumference of whose base is 14796
feet. Ans.

OF THE CONE.

246. *The convex surface of a cone is found by multiplying the circumference of its base by one-half its slant
height.**

*The solidity of a cone is found by multiplying the area
of its base by one-third of the altitude.*

Q. How is the convex surface of a cone found? The solidity? What
is the slant height of a cone? Altitude?

EXAMPLES.

1. Find the convex surface of a cone, whose slant height
is 50 feet, and the diameter of whose base is 10 feet.

We first find
the circumfer-
ence of the base
by Art. 243,

$3.1416 \times 10 = 31.416 =$ circle of base.
$31.416 \times 25 = 785.4 =$ convex surface.

and then multiply by 25, which is half the slant height.
The convex surface is 785.4 square feet.

---

* The slant height of a cone is the distance from its vertex to the circumference of its base. The altitude of a cone is the perpendicular distance from the vertex to its base.

2. Find the solidity of a cone, whose altitude is 24 feet, and the diameter of whose base is 12 feet.

$6 \times 6 \times 3.1416 = 113.0976 =$ area of base.
$113.0976 \times 8 \qquad = 904.7808 =$ solidity.

We first find the area of the base by multiplying the square of the radius by 3.1416 (Art. 243). Then multiply this area by 8, which is one-third of the altitude of the cone. The result is 904.7808 cubic feet.

3. Find the convex surface of a cone, whose slant height is 22 feet, and the radius of whose base is 3 feet.
Ans.

4. Find the convex surface of a cone, whose slant height is 5 yards, and the circumference of whose base is 1100 feet. Ans. 8250 sq. ft.

5. Find the solidity of a cone, whose altitude is 20 yards, and the diameter of whose base is 15 feet.
Ans.

6. Find the solid content of a cone, whose altitude is 32 feet, and the radius of whose base is 12 feet.
Ans. 14476.4928 cu. ft.

OF THE PYRAMID.

247. *The convex surface of a pyramid is found by multiplying the perimeter* of the base by half the slant height.*

The solidity of a pyramid is found by multiplying the area of the base by one-third the altitude.

Q. How do you find the convex surface of a pyramid? The solidity of a pyramid?

EXAMPLES.

1. What is the convex surface of a pyramid, whose slant height is 10 feet, and perimeter of its base 100 feet?

We multiply 100 feet by 5, which is half the slant height.

$100 \times 5 = 500$ sq. feet.

---

* The perimeter of the base is the sum of the sides which form the base.

L 2

2. Find the solidity of a pyramid, whose altitude is 18 feet, and the area of whose base is 100 square feet.

We multiply 100 square feet by 6 feet, which is one-third the given altitude.

$100 \times 6 = 600$ cubic feet.

3. Find the convex surface of a pyramid, whose slant height is 24 feet, and the perimeter of the base 100 yards.
Ans. 400 sq. yards.

4. Find the convex surface of a pyramid, whose slant height is 48 feet, and the sides which form the base measure respectively 8, 6, 5, 3, 2, 1 feet.
Ans.

5. Find the solidity of a pyramid, whose altitude is 9 feet, and the area of whose base is 25 square yards.
Ans. 25 cubic yards.

6. Find the solidity of a pyramid, whose altitude is 20 feet, and the area of whose base is 50 square yards.
Ans. 111 + cubic yds.

7. Find the solidity of a pyramid, whose altitude is 45 feet, and the area of whose base is 300 square feet.
Ans. 4500 cubic feet.

8. Find the solidity of a pyramid, whose altitude is 100 feet, and the area of whose base is 750 square feet.
Ans.

OF THE PRISM.

248. *The convex surface of a prism is found by multiplying the perimeter of its base by the altitude.*

The solidity of a prism is found by multiplying the area of its base by the altitude.

Q. What is the convex surface of a prism equal to? The solidity?

EXAMPLES.

1. Find the convex surface of a prism, whose altitude is 10 feet, and the sides of whose base measure 10, 5, 6, 7, 4 feet respectively.

We add the sides of the base together to get the perimeter, and multiply this sum by the altitude.

OPERATION.

$10+5+6+7+4 \times 10 = 320$ sq. ft.

2. Find the solidity of a prism, the area of its base being 100 square feet, and the altitude 5 feet.

We multiply 100, the area of the base, by 5, the altitude.

OPERATION.

$100 \times 5 = 500$ cubic ft.

3. Find the convex surface of a prism, whose altitude is 20 feet, and the perimeter of its base 200 feet.

Ans. 4000 sq. ft.

4. Find the convex surface of a prism, whose altitude is 150 feet, and the sides of its base, 50, 45, 40, 35, 30 feet respectively. Ans. 30000 sq. ft.

5. Find the solidity of a prism, whose altitude is 147 feet, and the area of whose base is 100 square yards.

Ans. 4900 cubic yds.

6. Find the solidity of a prism, whose altitude is 75 feet, and the area of whose base is 200 square feet.

Ans.

7. Find the solidity of a prism, whose altitude is 64 feet, and the area of whose base is 190 square yards.

Ans.

APPLICATIONS OF MENSURATION.

LAND SURVEYING.

249. Land is measured by a chain, 4 rods or 66 feet in length, called Gunter's chain, from the name of the inventor. This chain consists of 100 links; each link is therefore 7.92 inches. There are 80 chains = 320 rods in 1 mile.

Land is usually computed in *acres, roods, square rods* or *perches.* An acre of land is equal to 10 square chains; that is, to a rectangle 10 chains in length and 1 chain in breadth. There are therefore $40 \times 4 = 160$ square rods in 1 acre, and also $1000 \times 100 = 100000$ square links in 1 acre. Hence, in measuring the lines of a survey with the Gunter's chain, if we regard the chains and parts of chains as links, every chain being 100 links, the area will be expressed in *square links;* and since 100000 square links make 1 acre, the area will be reduced to acres and decimals of an acre, by cutting off 5 decimals from the right.

Again, since 40 square rods or perches make 1 rood, and 4 roods make 1 acre, we may reduce the decimals of an acre to roods and decimals of a rood by multiplying by 4; and the decimals of a rood are brought to square rods or perches by multiplying by 40.

Q. By what instrument is land measured? How many rods in Gunter's chain? How many links in a chain? What is the length of each link? How is land usually computed? How many square chains in an acre? What do you mean by saying an acre contains 10 square chains? How many square rods in an acre? How many square links in an acre? In computing the contents of land, are the lines reckoned as chains or links? Will the area be expressed in square chains or square links? How may it then be brought to *acres?* How do you bring the decimals of an acre to roods? Why multiply by 4? How do you bring the decimals of a rood to perches? Why multiply by 40?

To Survey a Triangular Field.

250. Having set up marks at the three angles of the triangle, measure with the chain the base A B, which we suppose to be 17 chains 8 links, or 1708 links, and then measure a line where the perpendicular from C would probably be; that is, C P=458 links. To compute the area, we multiply the base, 1708, by half the altitude =: 229 (Art. 241), and the area in square links is found to be 391132. Pointing off 5 places of decimals to the right brings the square links to acres (Art. 249), and the decimals of an acre are brought to roods by multiplying by 4, and the decimals of a rood to perches by multiplying by 40. The triangular field contains 3 A. 3 R. 2 P.+

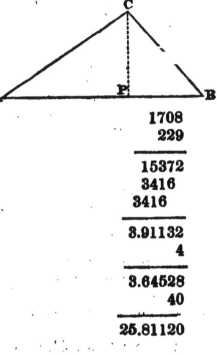

$$
\begin{array}{r}
1708 \\
229 \\
\hline
15372 \\
3416 \\
3416 \\
\hline
3.91132 \\
4 \\
\hline
3.64528 \\
40 \\
\hline
25.81120
\end{array}
$$

Area = 3 A. 3 R. 25 P.+

1. Find the content of a triangular field, whose base being measured, contains 15 chains and 27 links, and altitude 2 chains 13 links.   Ans. 1 A. 2 R. 20 P.+

2. Find the content of a triangular field, whose base measures 2915 links, and altitude 2 chains.
                                    Ans.

3. What is the area of a triangular field, whose base measures 25 chains, and altitude 95 links?
                                    Ans.

To Survey a four-sided Field.

251. Measure with the chain the line connecting two opposite angles of the field, as A C. The line A C will divide the quadrilateral into two triangles, whose common base may be taken to be A C. Then measuring the perpendiculars B F and D H, we may readily compute the

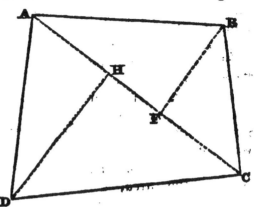

area of the two triangles A B C, A D C, and their sum will give the area of the whole field. The line A C is called a *diagonal.*

If the four-sided figure were a square or rectangle, we would measure the two adjacent sides, and their product would give the area sought.

If it were a rhombus or rhomboid, we would measure one side, and the altitude and their product would give the area.

If it were a trapezoid, we would measure the two sides which are parallel, and then the distance between the parallel sides, the sum of the parallel sides multiplied by half the altitude, would give the area.

EXAMPLES.

1. Find the content of a four-sided field, one of its diagonals being found to be 12 chains, and the altitudes of the two triangles formed being 4 chains and 2 chains respectively.

OPERATION.

$1200 \times 200 = 240000$ sq. links $= 2$ A. 1 R. 24 P. $+$
$1200 \times 100 = 120000$ " $= 1$ A. 0 R. 32 P. $+$

Ans. 3 A. 2 R. 16 P.

We find the content of each of the triangles into which the *diagonal* divides the field, and add the two areas together.

2. Find the area of a square field, whose sides measure 270 chains 4 links.   Ans. 7292 A. 0 R. 25 P.+

3. Find the area of a rectangular field, whose length is 140 chains, and width 75 chains 41 links.
                    Ans. 1055 A. 2 R. 38 P.+

4. What is the area of a rhomboidal field, whose base is 29 chains, and altitude 13 chains 2 links?
                    Ans.

5. Find the area of a trapezoidal field, whose parallel sides measure 23 chains and 14 chains respectively, and whose altitude is $3\frac{1}{2}$ chains. Ans.

To survey an irregular Field.

252. We fix marks at the most prominent points of the field, and then divide it into triangles and trapezoids, the triangles having their angular points, at the marks which have been made. Measure the bases and altitudes of each triangle, and the parallel bases and altitudes of the trapezoids; then compute the area of these figures as has been explained: the sum of the areas will be the area of the field.

Thus, find the area of the field A B C D, &c., the following lines being measured.

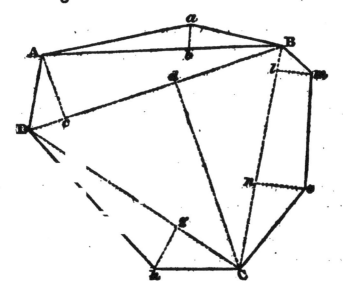

A B = 17 chains	8 links		ab.	= 25 links.
B C = 16 "	14 "		gh.	= 50 "
C D = 21 "	12 "		lm.	= 20 "
A D = 5 "	1 "		no.	= 55 "
D B = 19 "	11 "		Bl.	= 24 "
C d = 10 "	14 "		nc.	= 5 chains.
A c = 4 "	1 "			
nl = 4 "	50 "			

Ans. 14 A. 2 R. 11 P. +

Survey of the Public Lands.

253. In surveying the public lands, they are divided into squares, the sides of which are 6 miles, by a system of east and west lines intersecting another system of north and south lines called *meridians.* These squares are called *townships.* Each township, therefore, contains 36 square miles, called *sections.* The sections are again divided, in like manner, into half-sections, quarter-sections, and eighths of sections. From which it follows, that

1 township contains 36 sections=23040 acres.
1 section   "   640 acres.
1 half section "   320 "
1 quarter "   "   160 "
1 eighth   "   "   80 "

CARPENTER'S WORK.

254. Carpenter's work, such as roofing, flooring, partitioning, &c., is usually estimated by the square of 100 square feet; that is, a square whose sides are 10 feet.

To find the number of squares in a roof, &c., multiply the length by the breadth, and point off two decimals from the right.

EXAMPLES.

1. How many squares in a roof which is 50 feet long, and whose rafters measure 22 feet?

After finding the number of square feet in one side of the roof, we multiply by 2 for both sides. *Pointing off* two decimals, the answer is 22 squares.

OPERATION.

$22 \times 50 = 1100$
$1100 \times 2 = 2200$

Ans. 22.00 squares.

2. How many squares in a roof whose length is 100 feet, depth 45, and pitch of roof 10 feet?   Ans. 492.+.

3. How many squares in a conical roof, the diameter of its base being 25 feet, and altitude 8 feet?

           Ans. 5.81 +.

4. How many squares in a square roof, the sides of the square being 40 feet, and the pitch of the comb of the roof being 10 feet?   Ans.

5. How many squares in a floor 48 feet 6 inches long, 24 feet 3 inches broad?   Ans. 11.76 +.

6. How many squares in a floor 36 feet 3 inches long, and 16 feet 6 inches broad?   Ans. 5.98.

7. How many squares in a partition 75 feet long, 24 feet 6 inches high?   Ans.

8. How many squares in a partition 98 feet 6 inches long, 36 feet 3 inches high?   Ans.

BRICKLAYER'S WORK.

255. Brick work is usually estimated by the thousand brick. A wall 1 foot long, 1 foot high, and 1 brick thick, is estimated to contain 13 bricks. For the same length and height, and 1½ brick thick, it contains 18 bricks; and for a two brick wall, 24 bricks.

Hence, to find the number of bricks in a wall, we multiply the length of the wall by the height, and this product by 13 for a 1 brick wall, by 18 for a 1½ brick wall, and by 24 for a 2 brick wall.

An allowance is made for the windows and doors, by throwing out one-half their cubic dimensions. Chimneys are estimated as solid.

EXAMPLES.

1. Find the number of bricks required to build a house 40 feet square, 25 feet high, walls 2 bricks thick for the first story, 15 feet in height, and 1½ brick thick for the second story. Two partition walls, each 36 feet long, 25 feet high, and 1 brick thick. There are 4 stacks of chimneys,

each 8 feet wide, 3 feet deep, and 47 feet high. Four doors,
7½ by 4¾ feet, in the outside walls of the lower story; 10
windows in the lower story, each 6 by 4 feet; 14 windows
in the upper story, of the same dimensions as those in the
first story; 8 doors in the partition walls, each 7 by 3¼
feet. There are two gables, 1 brick thick, the pitch of roof
being 8 feet.

OPERATION.

$40+40+40+40=$　160 length of wall.
$160\times15\times24$　　$=57600$ bricks in lower story.
$160\times10\times18$　　　$=28800$ " " upper "
36×2　　　　$=$　72 length of partition.
$72\times25\times13$　　　$=23400$ bricks in partition.
$8\times47\times48\times 4=72192$ " " chimneys.
$7\frac{1}{2}\times4\frac{3}{4}\times24\times 4= 3240$ bricks in spaces for doors. ⎫
$6\times 4\times24\times10= 5760$ " " " windows. ⎬ 1st story.
$6\times 4\times18\times14= 5048$ " " " windows 2d st'y.
$7\times3\frac{1}{4}\times13\times 8= 2548$ " for doors in partitions.
$40\times 4\times13\times 2= 4160$ " in gables.

57600	3240
28800	5760
23400	5048
72192	2548
4160	
	2)16596
186152	
8298	8298

Ans. 177,854 bricks.

In computing the number of bricks in the chimneys, we
multiply by 48 for a wall 3 feet thick. After finding the
number of bricks that would be required, supposing there
were no windows and doors, we deduct half the bricks in-
cluded in windows and doors from the entire number found,
and the remainder gives the number of bricks required.

2. Find the number of bricks required in a house 48 by 24
feet, and 20 feet high, outside walls 2 bricks thick throughout, partitions 44 by 20; dimensions and details of house in
other respects same as in last example.　　　　Ans.

PLASTERER'S WORK.

256. Plasterer's work is estimated by the square yard. Deductions are made for doors, windows, and other openings.

EXAMPLES.

1. How many square yards of plastering in a ceiling 43 feet 3 inches long, 25 feet 6 inches broad?

Ans. 122¼.

2. How many square yards of plastering in a room whose length is 18 feet 6 inches, breadth 12 feet 3 inches, and height 10 feet 6 inches, allowing one-half for a door 7 feet 3 inches by 4 feet, and two windows 7 feet by 3 feet 6 inches? Ans.

3. Find the number of square yards of plastering in the house given in Example 1, Art. 255, both sides of partition walls being plastered, and allowance of one-half being made for openings. Ans.

4. Find the number of square yards of plastering in Example 2, Art. 255. Ans.

MASON'S WORK.

257. Mason's work is measured by the cubic foot, yard, or perch. The cubic measure being in all cases equal to the length, breadth, and depth of the work multiplied together.

EXAMPLES.

1. Find the solid contents in the stone wall of a cylindrical ice-house which is 18 feet deep, whose diameter from out to out is 24 feet, and the thickness of wall being 1 foot.

OPERATION.

$$12 \times 12 \times 3.1416 \times 18 = 8142.8472$$
$$11 \times 11 \times 3.1416 \times 18 = 6842.4048$$

Ans. 1300.4428 feet.

We first find the contents of a cylinder, the diameter of whose base is 24 feet (Art. 246), and from it subtract the content of a cylinder, the diameter of whose base is 22 feet. The result is the solid content of the wall.

2. Find the solid content of a wall 53 feet 6 inches long, 12 feet 3 inches high, and 2 feet thick.

Ans. 1310¾ feet.

3. Find the solid content of a wall, the length being 24 feet 3 inches, height 10 feet 9 inches, and 2 feet thick.

Ans. 521.37 + feet.

EXCAVATION.

258. The amount of excavation in canal work, trenching, &c., is estimated by the cubic yard or perch. When the section of the canal is of uniform dimensions, the content may be arrived at by multiplying the area of the section by the length of the canal. And in like manner we determine the amount of excavations in ditching. In general, multiply the length, breadth, and depth together for the cubic measure.

EXAMPLES.

1. Find the number of cubic yards in the excavation of a ditch which measures 4 feet wide at the top, 2 feet wide at the bottom, is 3 feet deep, and 20 yards long.

The section of the ditch is a trapezoid, the parallel sides of which are 4 feet and 2 feet, and altitude 3 feet. Its area is 9 square feet (Art. 242.) Multiplying 9 by 60 feet, the length of the ditch, the product 540 is the number of cubic feet in the ditch. This, divided by 27, gives the number of cubic yards, which we find to be 20.

OPERATION.

$\frac{(4+2)\times3}{2} =$ 9 square feet

$9 \times 60 = 540$ cubic feet.

$\frac{540}{27} =$ 20 cubic yds.

2. How many cubic rods in the excavation of a canal 100 miles in length, which is 22 feet wide at the bottom, 30 feet at the top, and 5 feet deep? Ans.

3. How many cubic yards in the excavation of a cylindrical ice-house which is 18 feet deep, and 24 feet diameter? Ans.

4. Find the cost of excavating the New York and Erie Canal, which is 363 miles long, 40 feet wide at the surface, 28 feet wide at the bottom, and 4 feet deep, at 25 cents per cubic yard. Ans.

5. What would be the cost of excavation of the Chesapeake and Ohio Canal, at 25 cents the cubic yard, its dimensions from Georgetown (D. C.) to Pittsburgh being as follows: Length, 341¼ miles; width at surface, 70 feet; at bottom, 50 feet; and 6 feet depth? Ans.

BOOK-KEEPING.

BOOK-KEEPING *is the art of keeping mercantile accounts.*

There are two kinds of book-keeping in common use, viz. *Single Entry* and *Double Entry.*

Book-keeping by single entry means that system in which there is but a single entry made of the transaction; and in book-keeping by double-entry, every entry is double; that is, has both a creditor and debtor. The method by single entry is used by those whose business is limited; but in large mercantile establishments in which great care and accuracy are required, the double entry system is alone used.

We propose to present a brief practical explanation of both these systems.

BOOK-KEEPING BY SINGLE ENTRY.

In book-keeping by single entry three books are generally used.

1. *Day-Book*, in which are entered, as they occur, the transactions of each day.

2. *Cash-Book*, which contains a transfer of the receipts and payments in *cash* recorded in the day-book, the cash account being credited by the payments made, and debited with the receipts.

3. *Leger*, which is a condensed register of all the entries in the day-book, so arranged as to exhibit on one side all the *sums* at debtor, and on the other all those at creditor.

DAY-BOOK.

Philadelphia, Nov. 1, 1844.

Page		$	cts.	$	cts.
	Thomas Smith Dr.				
	To 2 doz. slates at $4 per doz.	8			
	" 6 reams letter paper at 3\frac{50}{100}$.	21			
	" 6 " cap " " 3\frac{25}{100}$,	19	50		
	" 2000 quills at $10 per M. ..	20		68	50
	———— " ————				
	James Johnson & Co. Dr.				
	To 1 box tobacco, 100 lbs. at 10 cts.	10		10	
	———— 2 ————				
	Greer, Samuels, & Co. Dr.				
	To 1 roll carpeting, 65 yds. at 1$\frac{06}{100}$ per yd.	68	90		
	" 1 piece binding	1	50		
	" 1 rug	5	75	76	15
	———— 4 ————				
	Wm. Collins Dr.				
	To 1 horse	125			
	" 1 saddle and bridle	32	25	157	25
	———— " ————				
	Wm. Collins Cr.				
	By 100 bushels of wheat at 1\frac{05}{100}$	105			
	" cash in full of acco't.	52	25	157	25

Philadelphia, Nov. 4, 1844.

Page		$	cts.	$	cts.
	James W. Ross, *Dr.*				
	To 5 barrels flour at $6,$\frac{50}{100}$	32	50		
	" cash loaned	15	00	47	50
	"				
	Robert Thompson, *Cr.*				
	By 10 hogs weighing 2275 lbs. at 5½ cts.	113	75		
	" 1 quarter of beef, 105 lbs. at 8 cts.	8	40	122	15
	"				
	Thomas Smith, *Cr.*				
	By cash on acco't.			50	00
	5				
	James Johnson & Co. *Cr.*				
	By cash in full of acc't.			10	00
	6				
	Robert Thompson, *Dr.*				
	To cash on account	25			
	" 1 bbl. molasses, 31 gls. at 30 cts.	9	30		
	" 1 " sugar, 109 lbs. " 10 "	10	90	45	20
	7				
	Green, Samuels & Co. *Cr.*				
	By cash in full of acc't.			76	15

Philadelphia, Nov. 8, 1844.

Page			$	cts.	$	cts.
	James Alexander,	*Cr.*			70	00
	By 20 cords of wood at $3 50/100					
	"					
	James Alexander,	*Dr.*			70	00
	To cash in full of acc't.					
	"					
	Robert Thompson,	*Dr.*				
	To cash		30	00		
	" 1 barrel mackerel		7	25	37	25
	"					
	George Brown & Son,	*Cr.*				
	By 20 reams of paper at $4		80			
	" 15 boxes ink " $3 25/100 ..		48	75	128	75
	"					
	Robert Thompson,	*Dr.*				
	To cash		30	27		
	" 1 box raisins		2	25	32	52
	"					
	George Brown & Son,	*Dr.*				
	To 5 barrels flour at $6 50/100				32	50

It will be observed by noticing the entries made in the Day-Book, that each person is made Dr. to the articles purchased or cash paid, and Cr. by articles or cash received.

The numbers in the left-hand column indicate the number of the page in the Leger to which the entry is transferred.

LEGER.

INDEX.

A.	**N.**
Alexander, J.	**O.**
B.	**P.**
Brown, G. & Son,	**Q.**
C.	**R.**
Collins, Wm.	Ross, J. W.
D.	**S.**
E.	Smith, Thomas
F.	**T.**
G.	Thompson, R.
Green, Samuels, & Co.	**U.**
H.	**V.**
I. J.	**W.**
Johnson, James	**X.**
K.	**Y.**
L.	**Z.**
M.	

LEGER.

Page L

| Dr. | | | Thomas Smith. | | | | Cr. | | | |

1844.			$	cts.	1844.			$	$	cts.
Nov. 1.	To sundries ..	1	68	50	Nov. 4.	By cash	2		50	
						" balance ...			18	50
			68	50					68	50
" 30.	To balance		18	50						

M

Page 2.

Dr.　　　　*James Johnson & Co.*　　　　**Cr.**

1844. Nov. 1.	To sundries ..	1	$ 10	cts. 00	1844. Nov. 4.	By cash	2	19	00

Green, Samuels, & Co.

1844. Nov. 2.	To sundries ..	2	76	15	1844. Nov. 7.	By cash	3	76	15

Wm. Collins.

1844. Nov. 4.	To sundries ..	2	157	25	1844. Nov. 4.	By merchandise	2	105	00
					" "	By cash	2	52	25
			157	25				157	25

James W. Ross.

1844. Nov. 4.	To sundries ..	2	47	50					

Dr.   *Robert Thompson.*   **Cr.**

1844.			$	cts.	1844.			$	cts.
Nov. 6.	To sundries ..	3	45	20	Nov. 4.	By merchandise	3	122	15
" 8.	" " ..	3	37	25					
" 9.	" " ..	3	32	52					
" 30.	" balance ...		7	18					
			122	15				122	18
					Nov. 30.	By balance ...		7	18

James Alexander.

1844.					1844.				
Nov. 8.	To cash	3	70		Nov. 7.	By merchandise	3	70	

George Brown & Son.

1844.					1844.				
Nov. 9.	To sundries ..	3	32	50	Nov. 8.	By merchandise	3	128	75
" 30.	To balance ...		96	25					
			128	75				128	75
					Nov. 30.	By balance ...		96	25

In forming the Leger, we commence with the first charge in the Day-Book against *Thomas Smith.* Instead of debiting with each item specified in the Day-Book, we make him Dr. to " *sundries,*" carrying forward the whole amount, $68.50. The figure 1 before the amount shows the page of the Day-Book on which the items included in the debit are enumerated. On page 2 of the Day-Book the name of Thomas Smith again occurs in a credit of $50. This is placed on the Cr. side of the Leger. The operation of transferring the charges from the Day-Book to the Leger is called *posting*. At stated periods, say once a month, the accounts in the Leger are *balanced ;* that is, the difference between the Dr. and Cr. sides are taken in each account, the Dr. side being charged with this difference when the Dr. side is less, and the balance carried forward to the next month as a credit; or the reverse, should the Cr. side be less. Thus, in the account of *Thomas Smith*, the balance against him is $18.50. He is credited with this balance, by which means the account for the month is *balanced*, and a charge is entered against him for the next month of this balance.

CASH-BOOK.

Dr.				Cash.				Cr.		
1844.			$	cts.	1844.				$	cts.
Nov. 4.	To amt. rec. from W. Collins	2	52	25	Nov. 4.	By this amount loaned J. W. Ross ..	2	15		
" "	" " " " T. Smith	2	50		" 6.	" " " pd. R. Thompson	2	25		
" 5.	" " " " J. Johnson & Co.	2	10		" 8.	" " " pd. J. Alexander	3	70		
" 6.	" " " " G. Samuels & Co.	3	76	15	" "	" " " pd. R. Thompson	3	30		
					" "	" " " do. do.	2	30	27
					" 30.	" Balance		18	13
			188	40					188	40
Nov. 30.	To balance cash on hand		18	13						

The entries in the Cash-Book are also made from the Day-Book. We run over each entry in the Day-Book, and note whenever *Cash* is paid or received. The first cash entry is on page 2 of Day-Book, in which W. Collins is credited with $52.25. This being received, Cash is made Dr. to $52.25. The amount paid to J. W. Ross is carried into the Cash-Book, as a *credit* to Cash. The balance being struck as the Leger accounts, the amount of cash on hand, Nov. 30, is $18.13, with which Cash is accordingly charged.

BOOK-KEEPING BY DOUBLE ENTRY.

The number and kinds of books kept in Double-Entry Book-Keeping depend upon the extent and character of the business of the house.

The ordinary and most essential books in an extensive business-house are the following:

1. *Day-Book, or Blotter*, in which are entered, as they occur, the transactions of each day.

2. *Journal*, which is a condensed transcript of the Blotter, so arranged that every debit shall have its corresponding credit, and every credit its corresponding debit.

3. *Leger.* Corresponding with the Leger in single-entry book-keeping.

DAY-BOOK.

Philadelphia, January 1, 1844.

		$	cts.
	Commenced business with the following means:—		
✓	Cash $3000.00		
	Merchandise per inventory 2000.00		
	Note of Wm. Allen 1000.00	6000	
	James Armstrong, **Dr.**		
✓	To 20 barrels family flour, $6.50 $130.00		
	" 100 bushels corn .... 50 cts. 50.00		
	" 5 half barrels buckwheat, 3.50 17.50	197	50
	2		
	Bo't of Wm. Boush for cash;		
✓	100 bbls. flour at $5.50	550	

Philadelphia, Jan. 3, 1844.

		$	cts.
Bo't of Alex'r Jones on acc't,			
20 barrels herrings at 2\frac{50}{100}$ $50.00			
10 half barrels shad at $3 .... 30.00		80	00
"			
Bo't of Henry Williams,			
100 boxes raisins at 2\frac{25}{100}$ $225.00			
Gave him my note at 60 days		225	00
"			
Wm. Barker, **Dr.**			
To 50 barrels flour at $6 .... $300.00			
" 50 boxes raisins at 3$\frac{75}{100}$ 187.50		487	50
Rec'd his note at 60 days for $300.00			
" cash for balance 187.50		487	50
4			
Rec'd from W. Allen,			
Cash for his note in my hands		1000	
"			
Bank of America has discounted Wm. Barker's note for $300, 60 days,			
Rec'd cash $297.00			
Interest for 2 mo. at 6 per ct. 3.00		300	00
5			
Rec'd from Jas. Armstrong on acc't,			
Cash		100	00
"			
Bo't of J. Armstrong on acc't,			
1 bale cotton 300 lbs. at 6$\frac{1}{4}$ cts. $18.75			
20 barrels tar at 2\frac{25}{100}$ 45.00		63	75

Philadelphia, Jan. 6, 1844.

		$	cts.
	Paid Alex'r Jones on acc't,		
✓	Cash	50	00
	——————— 8 ———————		
	Alex'r Jones, *Dr.*		
	To 5 barrels flour at 6\frac{25}{100}$	31	25
	——————— " ———————		
✓	Inventory of merchandise on hand this day $2202.50		

The Day-Book commences with a statement of the means of the house, including cash, property on hand, and bills due; and then the transactions of each day are entered as they occur. The above exhibits the usual form of making these entries; and from this book a transfer is made to the Journal.

JOURNAL.

The fundamental rule in forming the Journal is, *that every debit should have a corresponding credit, and every credit its corresponding debit.* In forming a list of balances between the debtor and creditor sides, the sum of the balances will be equal. Thus, if a merchant purchase a lot of shoes from B., a *fictitious* account is formed called "Shoe Account;" and in the Journal entry, Shoe Account is made Dr. to B. for the amount purchased. Whereas, in single entry, B. would be credited with the amount of the shoes bought; but no one is made debtor, because the goods are not sold.

This principle has caused merchants to form two classes of accounts, viz., *personal* and *nominal.* Personal accounts are debited for payments, sales, &c., made to *persons,* and credited for amounts received from them. Nominal accounts include payments made and amounts received on certain fictitious accounts which are intended to represent the merchant himself, or his gains or losses. The principal of these are stock account, profit and loss account, interest account, cash account, merchandise account, bills payable account, bills receivable, bank account, &c.

On commencing business, stock account is credited by amount on hand, and debited for debts due others.

Profit and loss account is credited with the gains on particular accounts, and debited with the losses.

Interest account is credited with amounts received on discount and interest, and debited with discounts and interests allowed.

Cash account is debited with all money received and credited with all money paid.

Merchandise account is debtor for all goods purchased, and creditor for all goods sold.

Bills payable account is credited for all notes, bonds, &c., which are given or accepted and debited for those which are paid or disposed of.

Bills receivable account is debited for all notes, bonds, &c. which are received, and credited by all which are given or disposed of.

Bank account is debited for all payments or deposits made, and credited by checks, drafts, &c. drawn.

With these explanations we may now make the entries in the Journal as far as the Day-Book is made out.

JOURNAL.

Philadelphia, January 1, 1844.

			$	cts.
Sundries	*Dr. to*	*Stock.*		
Cash		$3000.00		
Merchandise		2000.00		
Bills receivable		1000.00	6000	00
	"			
James Armstrong		*Dr.*		
To Merchandise			197	50

Philadelphia, January 2, 1844.

		$	cts.
Merchandise *Dr.*			
To Cash		550	00
———— 3 ————			
Merchandise *Dr.*			
To Alexander Jones		80	00
———— " ————			
Merchandise *Dr.*			
To Bills Payable		225	00
———— 4 ————			
Sundries *Dr. to Merchandise*			
Bills Receivable $300.00			
Cash 187.50		487	50
———— " ————			
Cash *Dr.*			
To Bills Rec'ble		1000	00
———— " ————			
Sundries *Dr. to* *Bills Receivable*			
Cash $297.00			
Profit and Loss ....,... 3.00		300	00
———— 5 ————			
Cash *Dr.*			
To James Armstrong		100	00
———— " ————			
Merchandise *Dr.*			
To James Armstrong		63	75
———— 6 ————			
Alexander Jones *Dr.*			
To Cash		50	00
———— 8 ————			
Alexander Jones *Dr.*			
To Merchandise		31	25

M 2

LEGER.

The Leger is formed from the Journal in the same manner as in Single Entry Book-Keeping, the nominal accounts being opened and balanced as the personal accounts.

INDEX.

A.		L.	
Armstrong, J.	2.	M.	
B.		Merchandise	1.
Bills receivable	2.	N.	
" payable	2.	O.	
C.		P. Q.	
Cash	1.	Profit and Loss	3.
D.		R.	
E.		S.	
F.		Stock	1.
G.		T.	
H.		U. V.	
I. J		W. X	
Jones, A.	2.	Y. Z.	
K.			

LEGER.

Dr.			Stock.				Cr.		
1844. Jan. 8.	To balance ...	$ 6000	cts. 00	1844. Jan. 1.	By sundries ..	1	$ 6000	cts. 00
				" 8.	By balance ...		6000	00

Dr.   *Cash.*   *Cr.*

Page 2.

1845.				$	cts.	1845.				$	cts.
Jan. 1.	To stock	1		3000	00	Jan. 2.	By merchandise	1		550	00
" 3.	" merchandise			187	50	" 6.	" A. Jones ...			50	00
" 4.	" bills rec'ble .			297	00	" 8.	" balance			3984	50
" "	" do. do. ..			1000							
" 5.	" J. Armstrong			100							
				4584	50					4584	50
" 8.	" balance			3984	50						

Merchandise.

1845.						1845.					
Jan. 1.	To stock	1		2000	00	Jan. 1.	By J. Armstrong	1		197	50
" 2.	" cash			550	00	" 3.	" sundries ...			487	50
" 3.	" A. Jones ...			80	00	" 8.	" A. Jones ...	2		31	25
" "	" bills rec'ble .			225	00	" "	" balance			2202	50
" 5.	" J. Armstrong	2		63	75						
				2918	75					2918	75
" 8.	" balance			2202	50						

Bills Receivable.

1845.						1845.					
Jan. 1.	To stock	1		1000	00	Jan. 4.	By cash	1		1000	00
" "	" merchandise			300	00	" "	" sundries ..			300	00
				1300	00					1300	00

Dr.   *J. Armstrong.*   **Cr.**

Page 3.

1845.			$	cts.	1845.			$	
Jan. 7.	To merchandise	1	197	50	Jan. 5.	By cash	1	100	
					" "	" merchandise		63	75
					" 8	" balance		33	75
			197	50				197	50
" 8	" balance		33	75					

A. Jones.

1845.					1845.				
Jan. 8.	To cash	2	50	00	Jan. 3.	By merchandise	1	80	00
" "	" merchandise		31	25	" 8.	" balance		1	25
			81	25				81	25
" 8	" balance ...		1	25					

Bills Payable.

1845.					1845.				
Jan. 8.	To balance ...		225	00	Jan. 3.	By merchandise	1	225	00
					" 8.	" balance		225	00

Profit and Loss.

1845.					1845.				
Jan. 4.	To bills rec'ble	1	3	00	Jan. 8.	By balance ...		3	00
" 8.	" balance ...		3	00					

BALANCE-SHEET.

If the entries in Leger be correct, the sum of the credits should be equal to the sum of the debits.

		Dr.		Cr.	
		$	cts.	$	cts.
1	Stock			6000	00
1	Cash	3984	50		
1	Merchandise	2202	50		
2	Bills Receivable	0000	00		
2	J. Armstrong	33	75		
2	A. Jones	1	25		
2	Bills Payable			225	00
3	Profit and Loss	3	00		
		6225	00	6225	00

BUSINESS FORMS.

BILLS.

Va. Military Institute

      To Thomas, Cowperthwait & Co.   Dr.

1844.

Dec. 1. To am't. of acc't. rendered $452.72

"   " To 12 Anthon's Horace, at $1 .... 12.00

"   "   " 4 doz. slates, at $4 per doz. 16.00

                                  $480.72

Va. Military Institute

      Bo't, of Wiley & Putnam,

1844.

Dec. 24.   1 set of Encyclopedia Britannica, ... $150.00

        Rec'd. pay't,

                          WILEY & PUTNAM.

RECEIPTS.

Philadelphia, January 1, 1845. Rec'd. of T. H. Williams, One Thousand Dollars, in full of all demands.

$1000. FRS. H. SMITH.

Philadelphia, January 1, 1845. Received of F. R. Shunk, Esq., Five Hundred $\frac{25}{100}$ Dollars on account.

$500.25. THOMAS, COWPERTHWAIT & Co.

ORDERS.

THOMAS W. GILMER, ESQ.

Dr. Sir: Please pay to James M'Dowell or order, One Hundred Dollars, and place the same to my account.

Lexington, Jan. 1, 1845. J. T. L. PRESTON.

BANK CHECK.

$1000.

Cashier of the Farmers' Bank of Va., Pay to James W. Pegram or order One Thousand Dollars.

Dec. 18, 1844. RICHARD MORRIS.

DUE BILL.

Philadelphia, December 10, 1844. Due, on demand, to Francis H. Smith, Five Hundred Dollars, value received.

JAMES M'DOWELL.

PROMISSORY NOTE.

$1700.

Six months after date, I promise to pay Silas Wright, Esq., or order, Seventeen Hundred Dollars, value received.

LEWIS TIMBERLAKE.

New York, January 1, 1845.

$590.50.

Six months after date, we jointly and severally promise to pay James Wilson Five Hundred and Ninety $\frac{50}{100}$ Dollars, value received.

J. MARSDEN SMITH.

Philada. Jan. 1, 1845.　　WM. WOODBRIDGE.

DRAFTS.

$100.　　　　　　　　　*Philada. Jan. 1, 1845.*

At sight, pay to John Marshall, or order, One Hundred Dollars, value received, and charge the same to my account.

FRS. H. SMITH.

To BERNARD PRESTON, Esq.,
　Richmond.

$325.75.　　　　　　　*Lexington, Va. Jan. 1, 1845.*

Three days after sight, pay to order of Webb, Bacon & Co. at the Farmers' Bank of Va., Three Hundred and Twenty-five Dollars and Seventy-five cents, value received, and place the same to the account of the Virginia Military Institute.

FRS. H. SMITH.

JAMES HEATH, Esq.,
　Auditor of Va.

FRENCH SYSTEM OF WEIGHTS AND MEASURES.

THE French system of measures, established since the Revolution, is founded upon the measurement of a quadrant of a meridian. The ten-millionth part of this quadrant was taken as the *unit of length*, called *mètre*, all the other linear measures being multiples or sub-multiples of the metre, in *decimal* proportion. The metre corresponds to the 3.281 English feet.

The unit of *weight* is the *gramme*, which is a *cubic centimètre*, or 100th part of a metre of distilled water of the temperature of melting ice. It weighs 15.434 English Troy grains.

COMPARISON OF FRENCH AND ENGLISH WEIGHTS AND MEASURES.

DECIMAL SYSTEM.

Long Measures.

French.		English.
Millimètre	=	0.03937 inches
Centimètre..............	=	0.39371 "
Decimètre	=	3.93710 "
Mètre.................	=	39.37100 "
Decamètre	=	32.80916 feet
Hectomètre	=	328.09167 "
Kilomètre	=	1093.63890 yards
Myriamètre	=	10936.38900 "

Measures of Capacity.

Millitre	0.06103	cubic inches	
Centilitre	0.61028	"	"
Dècilitre	6.10280	"	"
Litre (cubic dècimètre) ...	61.02803	"	"
Decalitre	610.28028	"	"
Hectolitre	3.5317	"	feet
Kilolitre	35.3171	"	"
Myrialitre	353.17148	"	"

Solid Measures.

French.	English.
Décistre	3.5317 cubic feet
Stère (cubic mètre)	35.3174 " "
Décastre	353.1741 " "

Superficial Measures.

Centiare .... =	1.1960 square yards
Are (a square decamètre) .. =	119.6046 " "
Décare .... =	1196.0460 " "
Hectare	=11960.4604 " "

Weights.

Milligramme .... =	0.0154 grains Troy
Centigramme .... =	0.1543 " "
Décigramme .... =	1.5434 " "
Gramme .... =	15.4340 " "
Décagramme	=154.3402 " "
Hectogramme .... =	3.2134 ounces "
Kilogramme	=2 lbs. 8 oz. 3 dwt. 2 grs. "
Myriagramme	=26.795 lbs. "

ANCIENT WEIGHTS AND MEASURES.

Weights.

Attic obolus .... =	8.2 Eng. Troy grs.
" drachma .... =	54.6 " " "
Lesser mina	=3,892. " " "
Greater mina	=5,464. " " "
Talent = 60 minæ .... =	¼ English cwt.
Old Greek drachm .... =	146.5 Eng. Troy grs.
" " mina	=6,425. " " "
Egyptian mina	=8,326. " " "
Ptolemaic mina of Cleopatra.	=8,985. " " "
Alexandrian " " Dioscorides	=9,992. " " "
Roman denarius .... =	51.9 " " "
Ounce .... =	437.2 " " "
Pound of 10 oz.	=4,150. " " "
" " 12 oz.	=4,918. " " "

24

Scripture Measures of Length.

Digit	=	0.7425	inches Eng.
Palm	=	2.97	" "
Span	=	8.91	" "
Lesser cubit	=	1.485	feet Eng.
Sacred "	=	1.7325	" "
Fathom	=	2.31	yards "
Ezekiel's Reed	=	3.465	" "
Arabian Pole	=	4.62	" "
Schænus	=	46.2	" "
Stadium	=	231.	" "
Sabbath-day journey	= 1,155.	" "
Eastern mile	=	1.886	miles Eng.
Parasang	=	4.158	" "
Day's journey	=	33.264	" "

Grecian Measures of Length.

Dactylos	=	0.75546	Eng. inches
Doron } Dochme }	=	3.02187	" "
Lichas	=	7.55468	" "
Orthodoron	=	8.31015	" "
Spithame	=	9.06562	" "
Pous	=	1.00729	" feet
Pygme	=	1.13203	" "
Pygon	=	1.25911	" "
Orgyia	=	1.00729	" paces
Stadios } Dulos }	= 100.72916	" "
Milion	= 605.8333	" "

Roman Measures of Length.

Digitus transversus	=	0.72525	Eng. inches
Uncia	=	0.967	" "
Palmus minor	=	2.901	" "
Pes, the foot	=	11.604	" "
Palmipes	=	1.20875	" feet
Cubitus	=	1.4505	" "
Gradus	=	2.4175	" "
Passus	=	0.967	" paces
Stadium	= 120.875	" "
Milliare	= 967.	"

Roman Dry Measures.

Hemina .... = 0.5074 Eng. pints
Sextarius .... = 1.0148 " "
Modius .... = 1.0141 " pecks

Attic Dry Measures.

Xestes .... = 0.9903 Eng. pints
Chenix .... = 1.486 " "

Jewish Dry Measures.

Gachal .... = 0.1949 Eng. pints
Cab .... = 3.874 " "
Gomor .... = 7.0152 " "
Seah .... = 1.4615 " pecks
Ephah .... = 1.0961 " Winchester bu.
Coron .... = 1.3702 " quarter

Roman Measures for Liquids.

Hemina .... = 0.59759 Eng. pints
Sextarius .... = 1.19518 " "
Congius .... = 7.1712 " "
Urna .... = 3.5857 wine gallons
Amphora .... = 7.1712 " "
Caleus .... = 2.2766 hhds.

Attic Measures for Liquids.

Cotylus .... = 0.5742 Eng. pints
Xestes .... = 1.1483 " "
Chous .... = 6.8900 " "
Metretes .... = 10.3350 wine gallons

Jewish Measures for Liquids.

Caph .... = 0.8612 Eng. pints
Log .... = 1.1483 " "
Cab .... = 4.5933 " "
Hin .... = 1.7225 wine gall.
Seah .... = 3.4450 " "
Bath .... = 10.3350 " "
Coron .... = 1.6405 hhd.

Scripture Weights.

A shekel . . . . = 9 dwt. 2.6 grains
Maneh . . . . = 2 lb. 3 oz. 6 dwt. 10.3 gr.
Talent . . . . =113 lb. 10 oz. 1 dwt. 10.3 gr.

Scripture Money.

		cts. m.
Shekel of silver	=	53.3+
Bekah " "	=	26.6+
Zuza " "	=	13.3+
Gera " "	=	2.6
Maneh " "	=$26.65 cts.	
Talent " "	=$1599.	
Shekel " gold	=$8.65 cts. 9 m.+	
Talent " "	=$26,228.55 cts. 8 m.+	
Dram " "	=$5.19 cts. 8 m.+	

THE END.

Lightning Source UK Ltd.
Milton Keynes UK
UKHW021214041218
333417UK00020B/1839/P